T0214649

Brokering Circular Labour Migration

This book is about the global trend of commercialisation of domestic and care work through private agencies that organise transnational care arrangements by brokering migrant workers.

Situated in Switzerland, it focuses on the emergence of private home care agencies following the extension of the agreement on the free movement of workers to Eastern European countries in Switzerland. The agencies recruit migrant women from these countries and place them in private households for elderly care. This book explores how circular labour migration for these care workers is facilitated. In the form of a mobile ethnography, it traces their journey from Eastern European countries to Switzerland – from when care workers find employment and are recruited by agencies to when they arrive at their designated households. Taking care agencies as an analytical vantage point, it examines the care agencies' recruitment and placement practices and their role in facilitating migration.

Brokering Labour Migration offers an understanding of new migration patterns and highlights fundamental changes in migration control with the extension of free movement of workers in Switzerland to lower-wage countries in Eastern Europe. It will be an invaluable resource for academics and scholars of geography, anthropology, sociology, and gender and migration.

Huey Shy Chau is Affiliated Researcher in the Economic Geography Group at the University of Zurich, Switzerland. Her goal as a migration scholar and economic geographer is to explore how place and space matter and how they are shaped when people, knowledge, and social and economic practices circulate. She is driven by the motivation to understand social processes and transformation in relation to socio-economic inequality through the lens of mobilities, migration, and labour. What interests her most are the sites of struggle and negotiation around workers' and migrants' access to resources and social participation, freedom of movement, and the value of work.

Routledge International Studies of Women and Place

Series Editors

Janet Henshall Momsen and Janice Monk

University of California, Davis and University of Arizona, USA

For a full list of titles in this series, please visit www.routledge.com/series/SE0406

Brokering Circular Labour Migration

A Mobile Ethnography of Migrant Care Workers' Journey to Switzerland

Huey Shy Chau

Routledge
Taylor & Francis Group

LONDON AND NEW YORK

First published 2020
by Routledge
2 Park Square, Milton Park, Abingdon, Oxon OX14 4RN

and by Routledge
605 Third Avenue, New York, NY 10017

Routledge is an imprint of the Taylor & Francis Group, an informa business

First issued in paperback 2021

Publisher's Note
The publisher has gone to great lengths to ensure the quality of this reprint but points out that some imperfections in the original copies may be apparent.

British Library Cataloguing-in-Publication Data
A catalogue record for this book is available from the British Library

Library of Congress Cataloging-in-Publication Data
A catalog record for this book has been requested

ISBN 13: 978-0-367-14057-1 (hbk)
ISBN 13: 978-1-03-223812-8 (pbk)
ISBN 13: 978-0-429-02990-5 (ebk)

DOI: 10.4324/9780429029905

Typeset in Times New Roman
by Apex CoVantage, LLC

Contents

Tables and figures

Tables

Figures

Glossary

Betreuung	care
Pendelmigration	circular migration
Pro-Senectute	specialised agency and service organisation for age-related issues (*Schweizer Fach- und Dienstleistungsorganisation für Altersfragen*)
Respekt	care workers group supported by the union VPOD
Saisonniers	seasonal workers
Sans-papiers	undocumented migrants
Spitex	outpatient care services in Switzerland (*Spital-externe Dienste*)

Abbreviations

AMFP   Agreement on the Free Movement of Persons
DSL   Digital Subscriber Line
EFTA   European Free Trade Association
EU   European Union
EU-8   European Union accession states: Czech Republic, Slovakia, Hungary,
        Poland, Slovenia, Estonia, Latvia, and Lithuania
FIZ   Women's Information Centre
ICT   Information and Communications Technology
ILO   International Labour Organisation
OECD   Organisation for Economic Co-operation and Development
SECO   State Secretariat for Economic Affairs (*Staatssekretariat für Wirtschaft*)
STS   Science and Technology Studies
UNIA   Swiss labour union (*Die Gewerkschaft*)
VPOD   Swiss civil servant labour union (*Schweizerischer Verband des Personals
        öffentlicher Dienste*)
AOZ   Zurich Asylum Organisation

Currencies

CHF   Swiss Franc (CHF 1 equalled around USD 1.12 and EUR 0.81 in
       May 2014)
EUR   Euro (EUR 1 equalled around USD 1.37 and CHF 1.22 in May 2014)
HUF   Hungarian Forint (HUF 1 equalled around USD 0.004 and CHF 0.003 in
       May 2014)

Exchange rates of 15 May 2014 calculated according to the online currency
converter XE ('XE Currency Converter' 2017)

Acknowledgements

First and foremost, I would like to express my gratitude to my PhD advisor, Karin Schwiter, for her guidance in, patience for, and support of my career goals and, most importantly, for constantly challenging me to sharpen my arguments and my writing. Moreover, I would like to thank the members of my PhD committee, Christian Berndt, Susan Thieme, and Brenda Yeoh, for their generous support, invaluable advice, and encouraging words.

I found support and stimulation among my colleagues in the Department of Geography. The members of the Economic Geography Group – Johanna Herrigel, Simon Sontowski, Peter Latzke, Heidi Kaspar, Christin Bernhold, Leigh Johnson, Caroline Schurr, Manuel Wirth, Jennifer Steiner, Isabella Stingl – have given me constructive feedback on chapter drafts, which helped me develop the ideas presented here. Invaluable input also came from our reading group on intermediaries and brokers: Susan Thieme, Alice Kern, Yahel Ash Kurlander, and my project co-worker, Katharina Pelzelmayer. Thank you, Katharina, for many humorous moments in our shared office. Moreover, I cannot thank Roya Soleymani Kohler and the IT team enough for saving my data and providing support for technical issues.

I thank Shirlena Huang for facilitating a visiting fellowship in Singapore in spring 2016 and the members of the Social and Cultural Geography Seminar at the National University of Singapore for providing an inspiring environment for writing my dissertation. My appreciation also goes to Anju Paul, who put great effort into enabling my postdoctoral visiting fellowship at Yale–NUS College from January 2018 to June 2019 to work on a research project on domestic workers in Singapore. During this period, I received great input in revising my dissertation into the manuscript of this book.

Special thanks go to Katalin Rahel Turai for her wonderful collaboration out in the field. Her expertise in qualitative research and her sharp mind have been invaluable for my data collection. Daniela, Peter, and Andy deserve my thanks for their help in translating conversations, e-mails, and Facebook posts from Hungarian and Slovakian to English. I cannot thank Simon Milligan enough for his careful correction of my chapters and for finding better ways to convey what I wanted to say. I also thank Andrea Alleman, Manuel Wirth, Jasmine Truong, and Christin Bernhold for their help with earlier versions of some of the chapters. I especially thank my friend Corinne Wälti for her corrections and support in formatting my dissertation.

This research would not have been possible without the trust and hospitality of my informants. I sincerely thank my interview partners and the members of the Respekt group for sharing their time and experiences and for welcoming me into their homes and workplaces.

I would like to acknowledge with gratitude all my friends, in particular David Leutwyler, Ste, Andrea, Kri, Matt, Miri, Jane, Meli, Luki, Dane, Mare for their patience and support and for believing in me. I am forever grateful for my partner, Jeffrey Cheung. You are my everything. Thank you for cooking, cleaning, and caring when I was swallowed up by the deep sea of academic writing and for teaching me the importance of self-care after hours in front of my laptop.

And, finally, I am grateful for my siblings, Guo Lin and Guo Pyng, and their partners, Tham and Helen, whose love is with me in whatever I pursue. I am forever indebted to my parents for giving me the opportunities that have made me who I am. This journey would not have been possible if not for their unconditional support, and I dedicate this milestone to them.

1 The black box of live-in care labour migration

When Sara, a woman from eastern Slovakia who was unemployed at the time, saw a job advertisement on the social media platform Facebook, she spontaneously sent in her CV. An agency was looking for live-in carers to work in Switzerland. Shortly thereafter, a recruiter called Sara to offer her a temporary job: A man was recovering from a nervous breakdown and looking for someone to help with domestic chores. Sara would have to assist him in getting up in the morning and with keeping house. Two days later, she received a contract, which she signed and sent back to the agent. Seven days after that, Sara was in the back of an eight-seater car. Together with other care workers, she was on her way from Slovakia to Switzerland, where she would be dropped off at the care recipient's door.

The women started their journey in the evening and arrived at their workplaces the next morning. For some of them, it was their first time working as live-in care workers, while others were well acquainted with the job. Many of them were on their way to replace their 'switch partners', care workers with ending assignments. The women usually work for 2 to 12 weeks at a time before they are replaced by new or recurring care workers. After dropping off the new arrivals, the drivers pick up the departing care workers and drive back to Slovakia. These tours are arranged several times a month by the agency and are carried out in collaboration with a small transportation business in the recruitment country.

Sara's journey echoes the stories of many women around the globe who travel from poorer to richer regions to provide care work in private households. Care in the Global North is increasingly being delegated to women migrating from poorer to wealthier countries (Raghuram 2016). In Europe, there has been an increase in women from Eastern European countries working as care workers in wealthier EU countries. Initially, care workers worked mainly in informal arrangements and found work through informal networks (Krawietz 2014). Many started their work abroad by commuting to neighbouring countries, such as Germany and Austria. As early as in the 1990s, migrant workers had already established a system of going back and forth between their homes in Poland and their places of work in Germany (Irek 1998, 75; Morokvasic 1994; Schilliger 2014, 20). However, since the introduction of the Agreement on the Free Movement of Persons (AFMP) in Europe in 2002, there has been a rise in commercial agencies recruiting circularly migratin care workers from Eastern European countries. (Lutz 2008, 2011;

Triandafyllidou & Marchetti 2013). In Germany and Austria, the surge of live-in care for the elderly was particularly sparked by the introduction of the Posted Workers Directive, which allows agencies based in recruitment countries to provide services on a temporary basis (Bachinger 2009; Krawietz 2014; Österle, Hasl, & Bauer 2013; Rossow & Leiber 2017).

Switzerland differs from its neighbouring countries, as it does not allow the posting of workers. Moreover, it has gradually introduced the AFMP through bilateral agreements with the EU much later than EU member countries. Hence, discussions of live-in care by migrant care workers began almost a decade later than in the other German-speaking countries. The appearance of care agencies and live-in care was first documented in Switzerland in 2009, mainly in relation to informal arrangements (Schilliger 2009). However, the period since the extension of the AFMP to eight new EU member states in Eastern Europe in 2011 has seen a mushrooming of care agencies officially placing migrant care workers (Schilliger 2014; Truong, Schwiter, & Berndt 2012). In the last few years, these home care arrangements have become increasingly popular in Switzerland. More and more private households buy care services on a privatised care market. Private, for-profit care agencies not only broker migrant care workers to private households but also offer all-inclusive home care services. With their appearance, a new pattern of labour migration from households in Eastern Europe to households in Switzerland in the form of circular migration has emerged.

This development prompts many questions. How, and in which contexts, have these agencies developed, and how do they operate? How is this labour migration for live-in care work organised? How are social relationships created and unravelled in this process? In what ways do agencies control and shape labour migration within an open labour market that allows free movement of people between sending and recipient countries? Before we tackle these questions, this introductory chapter first presents the commercialisation of transnational live-in care as a global phenomenon and the existing literature on migrant domestic and care work. Subsequently, I outline the theoretical framework that underpins this research: migration infrastructure and politics of mobility. The third part presents the methods used for data collection and analysis. This introductory chapter finishes with an overview of the structure of this book.

Migration, gender, and care work

In the 1990s, feminist literature in the social sciences saw a rise in studies on an increasingly global phenomenon: migrant women leaving their own homes to work as domestic workers for other families (Momsen 1999; Constable 1997a; Huang & Yeoh 1996; Ehrenreich & Hochschild 2003). Women's increasing labour market participation in many countries in the Global North, which was not accompanied by an equivalent shift of care responsibilities to men, paved the way for hundreds of thousands of women to work as migrant domestic workers.

Unevenly gendered and global circulation of care

The migration patterns show that domestic work is mostly delegated to migrant women traveling from poorer to wealthier countries, leading to what has been discussed as an international division of social reproduction (Parrenas 2000). In addition to a gendered and global division in paid domestic and care work, scholars have observed a racial division, in which white women tend to occupy professional positions in the health care sector, such as certified nurses, whereas women of colour perform low-wage reproductive labour and are situated at the lower end of a racialised hierarchy (Glenn 1992; Parrenas 2001; Hondagneu-Sotelo 2000).

The many studies on the global rise of migrant domestic work tie into feminist discussions on unpaid domestic work and social reproduction in the 1970s. When Marxist feminist voices, in a context of international campaigns calling for wages for domestic work, started questioning a duality between women's' unpaid reproductive work ascribed to a private sphere and men's wage labour ascribed to a public sphere, they also challenged the implicit underlying gender contract that assigns care work to women (Cox & Federici 1976). Since the 1990s, feminist researchers have questioned how unpaid as well as paid domestic work is valued in society (see Kofman & Raghuram 2015 for an overview). The restructuring of unpaid care into paid care is closely related to the way economic concepts understand domestic and care work. How care is measured and defined in mainstream economic analysis has been questioned by a range of feminist economists (see Donath 2000; Folbre 1995; Knobloch 2009; Madörin 2007, 2010; Madörin & Soiland 2013; Perrons 2005; Waring & Steinem 1988). The literature has contributed to more gender inclusive economic studies by questioning the social construction of mainstream economic analysis that excludes the value of unpaid domestic work. Introducing the term 'the other economy', Donath (2000) argues that care work, which is concerned with the production of human beings, follows distinct economic logics that differ from the ones related to the production of commodities in other economic sectors. Similarly, Madörin emphasises that care work is interactive work (*personen bezogene Dienstleistung*) involving time and hence can only be rationalised to a certain extent (Madörin 2010). Care work differs from other types of work, as it encompasses not just manual but also affective embodied work and emotional labour (McDowell 2009). It is a kind of labour that 'requires one to induce or suppress feeling in order to sustain the outward countenance to produces the proper state of mind in others' (Hochschild 2003b [1983], 7; see also Steinberg & Figart 1999). It embodies attributes of a service that is usually financially unrewarded when 'undertaken by a close relation to the person cared-for' (McDowell 2009, 82). Because of its association with supposedly natural attributes of femininity, care work, according to McDowell (2009), is undervalued when the exchange takes place in the marketplace. Hence, wage payment is usually low considering the exhausting and difficult working conditions of live-in care.

In many countries, the household as a working place is characterised by a low level of labour regulations and relatively low salaries in comparison to other sectors (Chau, Pelzelmayer, & Schwiter 2018, 3–4; Medici 2012, 2015). Consequently, as Chen (2011, 170) points out, live-in domestic and care workers face greater isolation, limited mobility, and long working hours and are more exposed to abuse. Truong (2011) and Schilliger (2014) show that the live-in situations of care workers in Switzerland blur the boundaries between work and private life, which makes it difficult to distinguish the stipulated working hours from free time. Huang and Yeoh (2007, 212) show that abuse of domestic workers in Singapore can take place 'under the discourse of family' and is perpetrated by women employers on the basis of class asymmetries, ethnicity, and nationality. Moreover, they stress that uneven power relations in the household as a workplace and the characteristics of domestic work can hinder domestic workers' desires to leave employment and claim social justice in case of abuse (Huang & Yeoh 2007, 212; Yeoh, Huang, & Devasahayam 2004, 16). In contexts where domestic workers' only possibilities of accessing employment are through recruitment agencies and debt bondage, they are also more exposed to abuse from employers (Chen 2011, 170). Examples of such abuse include withheld wages and passports. In Europe, many care workers work in informal employment relations without access to social security; as a result, it is difficult for them to claim labour-related and social rights (Anderson 2000; Hess 2005; Karakayali 2010; Metz-Göckel, Münst, & Kałwa 2010).

Several studies underline the agency of domestic workers, showing how care workers employ strategies, negotiate work practices, and are active agents in shaping their own lives (England & Dyck 2012; Gaetano & Yeoh 2010; Pratt 2007; Rother 2017; Strüver 2011; Truong 2011; Yeoh & Huang 2010). Paul (2017) recognises domestic workers' agency by shedding light on how they mobilise social capital over the years and develop strategies of multi-step migration, working their way up to different destination countries. Furthermore, researchers address forms of political organisation of migrant workers, migrant organisations, human rights advocacy groups, and other non-governmental organisations advocating for improved working conditions and social rights (Ally 2005; Elias 2008; Piper 2005, 2007, 2010; Rehklau 2011; Rother 2009; Schilliger 2015). An important milestone was achieved when the International Labour Organisation (ILO) approved Convention 189, a document that stipulates labour protection around working conditions for domestic workers, after years of union and association organisation all over the world (Boris & Fish 2014; Fish & Boris 2015; Schwenken 2013).

As Williams (2011) emphasises, care is a global issue that requires global policy strategies to address the nature of unequal care distributions. She advocates for an understanding that recognises and redistributes care needs and care responsibilities, instead of seeing people as holders of individual rights, in order to address global justice issues. The unevenly gendered and global circulation of domestic and care work has been most prominently captured by the concept of global care chains (Ehrenreich & Hochschild 2003; Hochschild 2000; Parrenas 2000; Yeates 2004, 2012). It foregrounds that children and elderly family members of migrant care workers themselves are often in need of care, which would require the employment of migrant care workers from even poorer regions. Hochschild (2000, 2003a) described the phenomenon of a global care chain as an emotional imperialism, arguing that love as a resource is extracted from poorer countries to wealthier countries. While in earlier stages of imperialism, natural resources were expropriated from colonial countries, it would be love that is the new gold today. Love or emotional resources are quasi-extracted from poorer countries and accumulated in wealthier countries (Hochschild 2003a, 24, 27).

Focusing on bodily features of care work, Akalin (2015) proposes rereading the concept of global care chains in terms of affect production. According to Akalin (2015), the value of care labour is constituted through the fact that the affective labour produced by care workers is reserved for the employer's family, whereas the care workers' own families are denied the same level of care. Affect production, she argues, is immanent to the mobility of care workers. Hence, for Akalin (2015), affect labour foregrounds the fact that care work is relational, whereas in the concept of emotional labour, emotions occur only within the bodies of care workers at the moment of interaction. Similarly, Gutiérrez Rodriguez (2010) draws on concepts of affective labour to conceptualise the value of domestic and care work. Based on empirical data of Latin American migrant domestic workers and their employers in Europe, and by applying a postcolonial analytical lens, Gutiérrez Rodriguez (2010) presents the household as a space with transcultural encounters of unequal power relations and argues that migrant domestic work is more than just 'a field in which gendered and racialised boundaries are negotiated', that domestic workers engage with circuits of affective social (re)production (Gutiérrez Rodriguez 2010, 141). The amount of value that is produced with their labour is extracted and flows into the individual reproduction of the household, where the domestic workers are employed. For Akalin (2015), this extraction is only enabled by the fact that domestic workers are away from their own families, which are deprived of receiving the domestic workers' care. Crucial in Akalin's analysis is that a migrant domestic worker's 'history of affect making, i.e. whether she is married or not, how many children she has at home, how many siblings or relatives she has cared for', all play a role in the value creation of domestic work. From the point of view of employers, a domestic worker's association with motherhood embodies the capacity to produce ceaseless affection and the ability to respond aptly to any situations. Consequently, Akalin (2015, 74) argues that the 'left behind family is not an undesirable aspect or an unintended consequence of her mobility: it is an intrinsic aspect of how her labour gets exploited'.

The global flow of care has been further conceptualised as the care diamond (Raghuram 2012; Razavi 2007) and most recently as care circulation (Baldassar & Merla 2014). While the concept of a care diamond recognises that the provision of care involves multiple institutions, the concept of care circulation observes that care does not flow in only one direction but is rather multi-sited and occurs through asymmetrical reciprocal exchange within transnational families. Together, the literature on global migrant domestic and care work provides important insights into gendered and unequal distribution of care, on specific migration patterns, the often precarious transnational working and living conditions, the adverse impact on transnational families (Pratt 2012), and how care is reorganised in different societies. Many of these studies take individual domestic and care workers and/or the household as an analytical vantage point. The next sections present an overview of literature that takes brokers as a unit of analysis for migrant domestic and care work.

Recruitment agencies in domestic and care work

While many studies have mentioned the role of agencies as a subtopic, relatively few have taken them as a main unit of analysis in migrant care work. Some scholars have focused on the role of agencies in enabling employers to find carers and examined how agencies gain access to care workers. In a study in Los Angeles, Hondagneu-Sotelo showed that the agencies depict themselves as indispensable matchmakers to employers, who seek 'idiosyncratic traits, such as personal compatibility' (Hondagneu-Sotelo 1997, 5). Shedding light on a subsystem of local recruiters and informal intermediaries in Indonesia, Lindquist (2010, 2012) showed that agencies can play a key role in creating trust between would-be migrants and employers. Other studies stress the important role agencies play in facilitating complicated bureaucratic procedures to enable migration and in elevating standards for migrants (Goh, Wee & Yeoh 2017; Kern & Müller-Böker 2015). England and Dyck (2012) paid attention to the routes that outpatient home care workers, employed by home care agencies, take into care work in Canada. They observed that home care work 'was more readily available to them than other jobs because agencies were less concerned about their lack of work experience in Canada, gaps in their paid-work history, and, for the non-English speakers, their limited language skills'. In this sense, agencies were essential in providing employment to the outpatient care workers. As for live-in care work in Canada, a governmental programme enabled the hiring of migrant live-in care workers who, after a certain period of time, are eligible to apply for permanent immigrant status. In this context, Bakan and Stasiulis (1995) looked beyond the role of brokers for care workers' access to labour markets and emphasised the role of agencies as gatekeepers in negotiating citizenship rights for migrant workers in the Canadian context. In Europe, scholars have noted that care agencies often work with recruitment agencies in the recruitment countries to access care workers (Bachinger 2009; Krawietz 2014; Rossow & Leiber 2017). Another strategy is to encourage informal networks of carers to recommend other care workers to the agency (Schilliger 2014). Elrick and Lewandowska (2008) showed that many agents in Poland were deeply embedded in migrant networks; these authors argue that they play a significant role in continuing migration flows.

The literature also shows that domestic workers often have to pay a recruitment fee in order to gain access to employment. Shedding light on the recruitment process of Sri Lankan women into Middle Eastern households for domestic work, Eelens and Speckmann (1990) show that recruiters select those who bid the highest recruitment fees, whereas those who do not possess the financial means are denied access. As a consequence, many depend on moneylenders with high interest rates, which can lead to economic problems, particularly if domestic workers have to return prematurely from their employment. Moreover, the authors criticise illegal recruitment activities, such as fraud, for example when agents collect a fee but subsequently disappear, and the common practice of exceeding the stipulated maximum fees (Eelens & Speckmann, 1990). Hence, the recruitment of domestic workers can serve as viable business for many agencies. Lindquist (2012, 81) shows that for informal recruiters in Indonesia, it is even 'potentially far more lucrative to recruit migrants, particularly female domestic servants, than to become a migrant oneself'. Moreover, he points to a crucial difference in the recruitment between women domestic workers and male plantation workers; while men usually have to pay a fee before they depart, women domestic workers are often engaged in debt bondage. Similarly, Killias (2009) highlights that domestic workers from Indonesia are bound to their employers by placement and employment contracts that obligate them to pay back a debt. As a consequence, domestic workers are usually taken to training camps after they have agreed to work as domestic workers, where they are often confined for months before they leave for work (Killias 2009). The purpose in confining the domestic workers in these camps, as Lindquist (2010, 129) notes, is less the training of domestic workers but rather to prevent domestic workers from 'backing out' of their decisions. Moreover, Killias (2009) points out that debt bondage tends to prevent workers from leaving their employment, even in cases of abuse.

Another strand of literature focuses on production of stereotypes based on gender, nationalities, and other markers of identification fostered by employers and agencies. The stereotypes have been discussed as ethnicisation (Abrantes 2014; Bachinger 2009; Krawietz 2014; Schwiter, Berndt, & Schilling 2014) and racialisation (Bakan & Stasiulis 1995; Guevarra 2010; Hondagneu-Sotelo 2000; Liang 2011). Moreover, studies have shown that recruiters actively create divisions and hierarchies between workers. Hondagneu-Sotelo (2000) demonstrates that women from English-speaking countries are placed in the US as nannies, whereas women from Latin American countries are matched to households for domestic work. Similarly, Loveband (2004, 336) shows that Taiwanese agencies distinguish between domestic workers from Indonesia and the Philippines in ways that lead to Indonesian women often 'doing the dirtier and more demanding jobs'. Hence, the way that brokers promote essentialist stereotypes channels workers into specific sectors of the labour force. Moreover, while researchers examine the role of recruiters, they also consider the state in subjectivation processes of domestic workers into migrants and their role in producing an 'ideal' migrant worker in general in migrant-sending countries (Rodriguez 2010; Rodriguez & Schwenken 2013; Findlay et al. 2013; Shubin, Findlay, & McCollum 2014).

The traits that agencies seek and promote in care workers vary according to context. In their study on Sri Lankan domestic workers in the Middle East, Eelens and Speckmann (1990) observe that agents preferred to recruit women from rural places, as they are deemed to be more accustomed to hard work. In her examination on the recruitment, training, matching, and disciplining of live-in migrant care workers in Taiwan, Liang (2011) finds that agencies look for naïve, childlike, and innocent characteristics in applicants, who can be transformed into submissive and obedient workers in training centres. Similarly, researchers have shown how recruitment companies and non-governmental organisations in Indonesia (Killias 2009; Rudnyckyi 2004), and recruitment agencies and the state in the Philippines (Guevarra 2010), employ techniques, technologies, and (neoliberal) strategies to cultivate docile and passive domestic workers. Adopting a Foucauldian concept of governmentality, the authors show how these practices encourage domestic work-ers' self-governance in the production of the subject of docile and ideal domestic workers (Guevarra 2010; Rudnyckyi 2004). In Europe, scholars have found a very different idealised picture of live-in migrant care workers. Recruitment agen-cies portray care workers as devoted, family-oriented, and mature women from Eastern European countries with traditional values and the skills to adapt to their places of work (Bachinger 2009; Krawietz 2014; Schilliger 2014).

In sum, focusing on recruitment has allowed scholars to understand and 'recon-ceptualise gendered regimes of transnational migration' (Lindquist 2010, 117). Studies have shed light on the subjectification of domestic workers, the crea-tion of local and global divisions between domestic and care workers, and the role of agencies in facilitating access to employment. As Constable (2009, 54) summarises,

> a number of scholars have examined the role of domestic worker recruitment and employment agencies in marketing and selling products, distinguishing among different nationalities of workers, objectifying workers by offering specials, sales, markdowns, free replacements, and guarantees – in short, using the language of commodity markets to refer to workers.

Much of this research, however, presumes the existence of such labour markets and emphasises the power of the recruiters and agencies as gatekeepers. Less is known about the role of home care agencies in a shifting context from informal live-in care to the establishment of a formal labour market or about their role as key drivers of the commoditisation. Moreover, relatively little is known today about the transnational placement of live-in care workers and how control of their mobility and migration is constituted on an everyday level.

Opening the black box of migration in live-in care

As Lindquist, Xiang, and Yeoh (2012) note, in migration research, generally much more 'is known about why migrants leave home', while considerably less is known about how their mobility is made possible. This knowledge gap has been identified as 'the black box of migration' (Lindquist, Xiang, & Yeoh 2012). Notably, the role of labour migration brokers – people and institutions acting as intermediaries between would-be migrant workers and employers – has not been sufficiently examined. However, as long as we treat labour migrants' movements in isolation from the 'infrastructure', that is, 'the institutions, networks and peo-ple' (Lindquist, Xiang, & Yeoh 2012, 9) that facilitate their movement, society fails to consider labour migration as a socially constructed and human-driven phenomenon.

Moreover, a focus on how care workers are moved from their home country and the country wherein they are employed is key to understanding the mechanisms that control and shape their mobility and migration. Understanding how control of labour migration works is important, as it helps to unpack the underlying power relations that are inherent and (re)produce social and economic inequalities in the societies in which we live. These power relations are reflected among others in migrants' struggles to access labour markets in wealthier countries and in states' attempts to restrict and control migration. In our case, live-in care workers migrate in a context in which migration policies have shifted from national admission policies governed by the Swiss nation state as the main legal authority over movement to a policy of free movement of persons that is legally stipulated in supranational regulations. Consequently, attempts to control migration have arguably become more dispersed and less obvious to grasp. The question arises: In what way do newly emerged home care agencies play a role in this process? My aim is to open this 'black box of migration' (Lindquist, Xiang, & Yeoh 2012) and to shed light on the processes that facilitate live-in care workers' mobility and migration. Taking care agencies as an analytical vantage point, I trace the journey of migrant workers into live-in care work in Switzerland and examine the home care agencies' recruitment and placement practices. To do so, I draw on two conceptual approaches: politics of mobility and migration infrastructure.

Politics of mobility

The first conceptual foundation concerns a body of literature concerned with the mobility and movement of people, ideas, and things known as the 'new mobilities paradigm' (Cresswell 2006; Sheller & Urry 2006; Urry 2007). The 'mobility turn' arose in response to sedentarist theories in the social sciences; these theories consider a sedentary form of life tied to bounded places to be the norm (Sheller & Urry 2006). Instead of understanding stability and bounded places as basic units in social research, mobilities scholars focus on the movement of people and things. Hence, the mobility turn marks a shift within the social sciences from a 'metaphysics of fixity to a metaphysics of flow' (Cresswell 2006, 25). However, unlike debates focusing on deterritorialisation processes, which celebrate mobility and liquidity in a global world and so tend to neglect the importance of space and place in social sciences, the new paradigm does not abandon the notion of fixity altogether (Sheller & Urry 2006). Instead, fixity and movement form a dialectic relationship that underpins social life. Mobilities are enabled and supported by immobilities, or what Urry (2003) calls 'moorings'. Correspondingly, mobilities scholars often examine mobilities and movement in relation to notions of fixity, immobility, and moorings; they pay attention both to social and spatial relations and to infrastructures of mobilities, and hence foreground the fact that movement is never without context.

Recently, scholars have also called attention to the 'larger apparatus of power in which (these) relations of mobility are situated and governed' (Sheller 2016, 17) and outlined approaches to a politics of mobility (Adey 2006; Cresswell 2010). In politics, Cresswell (2010, 21) includes the 'social relations that involve the production and distribution of power'. Accordingly, a politics of mobility addresses the question of how mobilities are produced as well as how they create social relations. He sees mobility as a 'resource that is differentially accessed' and distributed, and he proposes thinking about who moves to where and how mobility is presented and embodied (Cresswell 2010, 21). Moreover, he considers six aspects of movement: why, how fast, in what rhythm, what route, how does it feel, and when and how does it stop (Cresswell 2010). For Cresswell (2010, 21), a politics of mobility matters, because mobility as a resource is distributed differentially and, hence, is an important explanation in the production of 'some of the starkest differences'.

The focus on mobilities in care work has increasingly gained attention, as shown by the number of studies in home care, some more loosely related to the new mobilities paradigm (Baldassar & Merla 2014; Cuban 2013; Cuban & Fowler 2012; Huang, Thang, & Toyota 2012; Schwiter, Berndt, & Schilling 2014). The questions of power relations and mobility, however, are nothing new; they had a long tradition in feminist political economy approaches to social reproduction and in feminist geographers' research into gender and work long before the mobility turn (Cox 2006; Hanson & Pratt 1995; Mahler & Pessar 2001; Massey 1993). Most famously, in her essay on 'a global sense of place', Doreen Massey (1993) answered David Harvey's (1989) understanding of time–space compression, a term that refers to a shrinking distance between places due to information and communication technologies, which according to Harvey is driven by economic developments. By stating that 'time–space compression has not been happening for everyone in all spheres of activity', Massey (1993, 60) points out that there are other aspects, such as race and gender and not just capitalism and its development, which determine an understanding and experience of time–space compression. Moreover, she calls for a need to pay attention to the social differentiation of this experience. With her formulation of a 'power-geometry', she suggests taking into account the uneven and unequal positioning of people in relation to 'flows and interconnections' (Massey 1993, 61). She points out that individuals have different access to mobility, and some are more in control of it than others. Those in charge of their mobility and time–space compression may use it to increase their own power and social position, which can actively weaken others. Thus, differentiations in mobility, according to Massey, reflect and reinforce existing power relations and social inequalities (Massey 1993). Scholars have recently called for a framework that more explicitly unites the mobilities concept with research on movement between places in relation to gender, work, and power relations (Cresswell, Dorow, & Roseman 2016; Dorow & Mandizadza 2018; Dorow, Roseman, & Cresswell 2017; Roseman, Barber, & Neis 2015). Termed employment-related geographical mobility, this concept calls for 'looking at the way in which forms of mobility are either enabled by, or impeded by, other forms of (im)mobility' across different scales and for understanding the connections between them (Cresswell, Dorow, & Roseman 2016, 1792).

Adopting an understanding of politics of mobility is useful for understanding the differential experiences of mobilities of live-in care workers on an everyday level and for situating these experiences in a larger context of power relations and control of mobility and migration (Chau 2019). By taking into consideration who the agencies select as carers, how quickly they are employed, in what rhythm they travel back and forth between their own home and workplace, what route and transportation form they take, as well as when and under which circumstances their employments end, a politics of mobility can point to inequalities in the care arrangements of migrant care workers. In this sense, this book provides an ethnographic study of the (im)mobilities that constitute care workers' migration and sheds light on the politics of mobility in the organisation of live-in care arrangements.

Migration infrastructure

The second conceptual approach that underpins this research stems from an infrastructural understanding in exploring migration. Infrastructure is increasingly becoming an object of study and a lens of analysis in the humanities and social sciences. Scholars across science and technology studies (STS), sociology, geography, and anthropology contribute to a diverse debate on what infrastructure is and how its study can enrich our understanding of social life (see i.a. Larkin 2013; Star 1999; Amin 2014; Harvey, Jensen, & Morita 2017; Lockrem & Lugo 2012). Whereas researchers in critical political economy have treated infrastructure as a materialisation of abstract forces and capitalist reproduction, STS scholars have analysed infrastructures as networks and socio-technical constellations consisting of human and non-human agents that are an essential part of social life and its production (see De Coss-Corzo 2016).

The latter approach to infrastructure has recently made its way into migration studies. Calling for an 'opening of the black box' of migration, a notion used to examine the internal workings of a system in science and technology studies, Lindquist, Xiang, and Yeoh (2012) propose a focus on 'the middle space of migration', that is, on brokers and infrastructures that facilitate movement (see also Lin et al. 2017). Infrastructure here is understood as 'matter that enable[s] the movement of other matter' (Larkin 2013, 329). Based on rich ethnographic work in Asia, Xiang and Lindquist introduce the concept of 'migration infrastructure' (Xiang & Lindquist 2014; Lindquist & Xiang 2018). The authors show that low-skilled labour migration from Indonesia and China is now more intensively mediated than in the past. Instead of a line between two places, they propose to imagine migration as a space of mediation that is occupied by commercial recruitment intermediaries, migrants, technologies, bureaucrats, and non-governmental organisations (Xiang & Lindquist 2014, 142). Accordingly, 'it is not migrants who migrate, but rather constellations consisting of migrants and non-migrants, of human and non-human actors' (Xiang & Lindquist 2014, 124). With their proposed concept of migration infrastructure, that is, 'the systematically inter-linked technologies, institutions, and actors that facilitate and conditions mobil-ity', the authors aim at an unpacking of these spaces of mediation (Xiang & Lindquist 2014, 124).

The authors foreground five dimensions of migration infrastructure, which are not supposed to be understood as clear-cut domains but rather indicate distinct logics of operation in each dimension: the social (migrant networks), the humani-tarian (NGOs and international organisations), the technological (communication and transport), the regulatory (state apparatus and procedures for documentation, licensing, training, and other purposes), and the commercial (recruitment interme-diaries) (Xiang & Lindquist 2014, 124). By shedding light on how the growth of one dimension has led to the growth of another in labour migration infrastructure in China and Indonesia, the authors observe that this interplay has become self-perpetuating and self-serving in the last two decades. In other words, it has devel-oped its own dynamic. The interplay between the dimensions becomes a crucial force in the conditioning of migration flows. In Xiang's and Lindquist's case, the growth of migration infrastructure, especially the regulatory and commercial infrastructure, has been intensive rather than extensive. The development of the migration infrastructure has not broadened the scope of destinations and thus generated new capacities for migration. Instead, labour migration has become more cumbersome. For example, in order to migrate, would-be migrants have to undergo increasingly detailed medical check-ups stipulated by state policies and carried out by recruitment companies. Xiang and Lindquist (2014, 136) call this development 'infrastructural involution'.

The focus of migration infrastructure lies in the operational logic of how migra-tion is constituted in different contexts; the concept is not meant to be compre-hensive. The five dimensions are entangled with each other as they 'collide with and contradict one another' (Xiang & Lindquist 2014, 124). But in each dimen-sion, the defining modus operandi and driving forces differ. For the analysis of migration infrastructures, the authors propose examining migration as quotidian and processual operations, paying attention to the relational dimensions across the five dimensions, and understanding infrastructure as socio-technical constel-lations (Xiang & Lindquist 2014, 143).

Recently, Xiang, Lindquist, and Yeoh, in collaboration with Weiqiang Lin, dedicated a second special issue on 'migration infrastructures and the production of migrant mobilities' (Lin et al. 2017). The issue reconceptualises the infrastruc-tural understanding of Xiang and Lindquist's (2014) concept by linking it to an infrastructural approach of mobilities and the mobilities paradigm. Correspond-ingly, migration as a 'complex system of mobilities/immobilities' or as a 'bundle of mobilities' is understood as 'a product of infrastructures' (Lin et al. 2017, 2). Consequently, the ways 'infrastructures produce and mobilise migrant subjects thus have direct consequences on the way societies are being (re)organised through the resultant mobilities' (Lin et al. 2017, 4).

In light of the recent emergence of home care agencies and an intensification of labour intermediation of migrant care workers, I consider a framework that conceptualises actors and infrastructures that facilitate movements across borders a fruitful way to explore the phenomenon of migrant care workers and the growing live-in care market. The strength of the theoretical approach is twofold. Firstly, the notion of 'migration infrastructure' allows a focus on social and quotidian processes that organise and channel the migration of live-in care workers. By paying attention to social connections that facilitate the migration of care workers, we unpack the middle space of live-in care labour migration. Moreover, the concept of migration infrastructure was developed based on empirical data on migration in Asia. By applying it to the migration patterns of migrant live-in care workers in Europe, we also explore how dynamics of migration differ in Europe from Asia. Secondly, a migration infrastructural view questions the picture of migration flows by underlining the social and material infrastructures needed for migration. Instead of understanding migration primarily as a reaction to economic structures that supposedly require state regulations to control labour migration flows, it shows how live-in care labour migration requires work and constant efforts to build networks and to maintain relations between people and institutions to facilitate migration in order to organise the journey of live-in care workers. Hence, this book provides an ethnographic account of how migration infrastructures of home care workers come into being and how these infrastructures work.

Clarifications on usages of the terms migration and mobility

While at times it may seem that I use the terms mobility and migration synonymously, they are not the same. Migration here is understood as a 'bundle of mobilities' (Lin et al. 2017, 2), which I consider useful to link the infrastructural and 'new mobilities' approaches on which I base my research. Consequently, I use the term mobility to refer to sub-aspects of migration, such as when referencing the cycles of shifts or how fast a care worker is able to begin an employment. The whole and repeated movement from home to workplace of care workers and the bundle of immobilities in the form of infrastructure that enable these movements, then, is what I understand as migration.

I find this important to point out as, by consciously using the term migration, I am also taking a political stance. Consider the following: In her analysis of traditional conceptions of the terms migration and mobility and their developments in European Union (EU) policy documentations, Bauloz (2016, 3) argues that the terms have been 'conceptualised differently with regard to third-country nationals [non-EU and non-EFTA nationals] and citizens of the European Union'. While the movement of third-country nationals is referred to as migration, the movement of EU nationals within the European Union and European Free Trade Association (EFTA) area is referred to as mobility (Bauloz 2016, 3). Moreover, she observes development of a migration–mobility nexus in the policy papers. She demonstrates that movement considered as desirable because of its supposedly positive impact over the economy is increasingly reconceptualised as mobility. This includes, for example, circular migration within the EU and temporary, limited migration of third-state nationals. The purpose, as Bauloz (2016, 19) argues, seems to be to disassociate these forms of movement from the supposedly 'negative connotation[s] it has acquired following years of restrictive measures to manage migration in the Union'. In this sense, the term migration is supposedly associated with undesirable movements whereas mobility is associated with economically useful ones.

Therefore, I actively choose to use the term migration rather than labour mobility to refer to the repeated movement of live-in care workers from their home to their workplaces in Switzerland, primarily to oppose a construction between desirable and undesirable movements along lines of economic utility. Moreover, in so doing, I intend to evoke an analogy between contemporary circular migration in Europe and the temporary labour migration programmes known as 'guest-workers' programmes in the post-war period. The purpose is to emphasise that both forms of migration, contemporary circular migration in the EU and migration in the framework of guest-worker programmes, face similar protection problems regarding living and working conditions (see also Chau, Pelzelmayer, & Schwiter 2018; Wickramasekara 2011).

The scope of this mobile ethnography

When I started my research in 2013, I was lucky to be able to build on previous work. Jasmine Truong and Linda Schilling, former master's students in geography at the University of Zurich, worked on the topic in their master's thesis. Additionally, Jasmine Truong, Karin Schwiter, and Christian Berndt (2012) already published a study on the landscape of care agencies in the canton of Zurich. In the course of their work, they contacted around 33 care agencies and interviewed 16 care agencies; this allowed me access to the transcripts on which to build my own research. I complemented that list with new home care agencies that I came across during my online search for home care agencies. Using keywords such as 24-hour care (*24 Stunden Betreuung*), elderly care (*Seniorenbetreuung*), and home care (*Betreuung zu Hause*), I looked for care agencies offering live-in care services in Switzerland. The difficulty with researching home care agencies is that there is no central register that would enlist them. Furthermore, some of the agencies offer not just live-in care services for the elderly but also additional services such as hourly outpatient care or even temporary jobs in other sectors. In addition, there are many care agencies that do not specify which places they mediate. Consequently, it was difficult to sort out the ones that place care workers in Switzerland and not only in Germany or Austria. I only included agencies in my list that would explicitly indicate on their websites that they place care workers in Switzerland. In the period from autumn 2013 to summer 2016, I came across a total of around 75 home care agencies including the ones that Jasmine Truong and Linda Schilling originally contacted.

Access to the field and data collection

Between May 2014 and March 2015, I contacted 24 home care agencies in Switzerland. In addition, I contacted care workers through the Internet on online forums and social media, and care agencies abroad, mostly via e-mail. The barrier to gaining access to interview partners outside of Switzerland was much higher than within Switzerland. I felt very relieved when I acquired temporary help from Katalin Rahel Turai, a PhD student at the Central European University in Budapest, who speaks Hungarian and offered to work as a field associate for our project. Katalin contacted seven agencies and several care workers through the Internet. On betreut.ch, an online platform for children, elderly, and animal care, she searched for elderly care workers' advertisements by city and browsed them according to names, which she could identify as Hungarian. Through these efforts, we were able to meet care worker Anna, who we visited at her place of work in Switzerland twice. Anna later invited us to visit her in her home village in Hungary in April 2015, where we met her husband and son, her cousin Sybille, and her friend Kati, who both work as care workers. Moreover, she introduced us to the mayor of the village, with whom we had an informal conversation at his office. She also took us to a local café where a group of nine care workers met up for coffee. The care workers were happy to engage in questions about the organisation of live-in care work and agreed to let us record the whole conversation. The conversation with the mayor and the group conversation were later translated and transcribed by Katalin.

I came to know about another care worker from Hungary with the help of a fellow PhD student in my department, who met her on a bus from Hungary to Germany and Switzerland. As he knew that I was working on care workers, he asked if he could pass on her contact information to me. However, because of language barriers, I was only able to conduct the interview with the help of Katalin, who translated the conversation after my arrival in Budapest. Marina then gave us the contact information of her friends, Berta and Talia, who work as care workers in Germany and in Switzerland. Subsequently, we video chatted with Talia and visited Berta in Kaposvár (not anonymised), a city in southwest Hungary.

In the end, I conducted **interviews with 20 representatives of 13 different care agencies**. The sample includes a wide range of agencies, from one-person businesses with around ten care recipients to well-developed agencies with more than 100 care recipients. The sample also covers a range of price segments from CHF 2000 per month up to more than CHF 12,000 per month. One of the care agencies was located in Hungary and one was in Slovakia. The remaining care agencies were based in Switzerland. Additionally, I participated in a conversation with my advisor and colleague, Karin Schwiter, and a representative from the non-governmental organisation Caritas, which mediates live-in care workers from Romania and Slovakia. Out of all the interviews, I gained a somewhat deeper insight into the workings of two care agencies: the ones managed by Mattea, based in Slovakia, and Pascal, based in Switzerland. Mattea had invited me to visit her care agency in Slovakia, where, with the help of a translator that Katalin helped to organise, I could also talk to her employees. Moreover, I had the chance to accompany her and other care workers on a car drive from Slovakia to Switzerland and accompany three care workers when they arrived at the household in Switzerland to start their employment. Katalin and I also visited a group of care workers in a village in southern Hungary, where Pascal recruits his agency's care workers. After Pascal announced our visit, the local language teacher Irina, who coordinated the assignments of the care workers was friendly enough to organise interviews with five care workers at the German minority house, which also served as our accommodation. She also arranged a tour at the local history museum and a dinner with the president of the German minority association for us. Overall, we left the village with the impression that Irina played a significant role in the village community as a gatekeeper for would-be care workers.

In addition to the interviews with care agencies, I conducted **interviews with 13 care workers**. Of the 13 care workers, one was a would-be care worker, as she was preparing for but had not yet worked as a care worker; two had been working in Germany as care workers, but had tried and wished to find employment in Switzerland. Moreover, I intermittently visited the group Respekt. The group meets once a month on a Sunday afternoon to discuss issues regarding the everyday life of live-in care work and problems that arise from it. On one occasion, I met with the unionist of the Respekt group along with one of the care workers I had interviewed, who told me about her worries and problems in relation to her working conditions and administrative work.

Finally, the more formal interviews with the care agencies and care workers are complemented with informal conversations with care workers and unionists. In addition, interviews were conducted with a care recipient, a unionist, and a representative of the State Secretariat for Economic Affairs, to gain context about the recruitment of live-in care workers and their access to the live-in care labour market. During the interviews, I gathered copies of documents such as working contracts, care agency information leaflets and advertisements, information for care workers from care agencies on working conditions, samples of pay slips, CVs, and video presentations of care workers. Therefore, my material consists of recorded and transcribed interview data, informal conversations, field notes, and documents related to the marketing and working conditions of live-in care work.

The interviews were conducted in German (21), Hungarian (9), and Slovakian (6). The non-German interviews were translated into English by my field collaborator Katalin for Hungarian interviews and another translator for Slovakian interviews. Of the 36 interviews, 20 were fully transcribed verbatim, and 7 were partially transcribed. I use pseudonyms (fictive names) for all interviewees, and I anonymised them so that links cannot be traced between them and specific business practices. I have purposely modified my writing so that it becomes difficult for the reader to follow the specific practices of the different agencies throughout the chapters. Initially, I had given pseudonyms to the care agencies so that it was possible to understand the relations between care agent, care workers, and specific care agencies. However, because of this transparency, there was a danger that readers who are familiar with the field could identify the care agencies and trace their business practices throughout the chapters. My specific concern was that care agents and other individuals could use my dissertation to read about the business practices of other care agencies and that care agents could identify care workers that were working for their agency and come to know what they had told me. Hence, I deleted the pseudonyms of the care agencies and instead introduced phrases such as 'one of the care agencies'. Moreover, I anonymised my interview partners such that their relations to each other are not always recognisable. In doing so, it has unfortunately become more difficult for the reader to relate to the interview partners' views and stories, as the practices and stories have become much more fragmented. However, the decision to omit the links between some of my interview partners and between certain business practices and care agencies allowed me to tell their stories and to give insight into recruitment and placement practices without harming my interview partners.

Data analysis

I used grounded theory methods according to Charmaz (2014) to analyse the material. Drawing on Strauss and Corbin (1990), Charmaz proposes two main coding phases, initial and focused coding, before writing up the results. Initial coding moves quickly through a large amount of data from selected interviews and stays close to the data. In contrast, the purpose of focused coding is to establish first analytic directions by grouping codes into categories, which are then applied to the rest of the material (Charmaz 2014). Accordingly, I applied initial coding to full transcripts of eight interviews with care agencies, a group conversation with carers in Hungary, and three interviews with carers using the MaxQda analysis program. Subsequently, I grouped the codes into categories and used these to sift through the rest of the material. While doing so, I added new codes to categories, changed and shifted codes, and renamed and merged codes into subcategories. The subchapters in this book correspond to key categories relevant to the recruitment and placement of care workers. The different steps and phases are by no means distinctive in a clear-cut way. Rather, the different steps of coding and categorising, of writing down of descriptive results, and of putting categories into relation to each other overlap temporally. I do not claim to have adopted grounded theory in a strict way. Grounded theory is known for its purpose of generating theories, which has never been my primary goal. Instead, I used grounded theory to produce as much openness as possible in the beginning of the analysis, to break open texts, and to find links between stories.

Mobile ethnography

The material in this book is partially linked by ideas on mobile ethnography (Büscher & Urry 2009; Faist 2012). The concept addresses the spatial movement of people through direct observation of associated social practices. 'Through investigations of movement, blocked movement, potential movement and mobility, dwelling and place-making', Büscher, Urry, and Witchger (2011, 2) argue that social researchers can show 'how various kinds of "moves" make social and material realities'. Since a mobile method starts with mobility and migration patterns and not people or state regulations, it is also a great approach for avoiding methodological challenges in contemporary migration studies, such as methodological nationalism (Wimmer & Glick Schiller 2002), and essentialist assumptions on groups by hastily adopting an ethnic and national lens (Faist 2012).

Novoa (2015, 99, 105) defines a mobile ethnography 'as a translation of traditional participant observation onto contexts of mobility' but also emphasises that it might prove useful to complete observant participation with other techniques, such as the analysis of photos, videos, and interviews, to understand the interview partners' views. Although my main data consists not of participant observations in a traditional sense but rather of interviews that are complemented by observations and other forms of data, I do not see this as problematic. As Madden (2010, 77) states, 'talking with people, being with people, and observing people are not divisible ethnographic actions'. Instead, 'ethnographers talk, participate and observe simultaneously and the sum total of all these actions creates participant observation in its broadest sense' (Madden 2010, 77). Therefore, I consider the outcome of this book a mobile ethnography in that I focus on the practices of migrant live-in care workers, care agents, and other actors that result in mobility and immobility. I pay attention to the 'emplacement of migrant subjects' while examining the specific contexts in which actors move and act, as suggested by Smith (2005, 244). I chose stories and moments that I consider relevant for the mobility of migrant care workers and that give insight into the emergence of the migration infrastructures that facilitate and characterise live-in care labour migration.

The stories and situations that I chose to write about are not meant to be complete but are fragments. In line with my methodological stance on social constructivism, I see my role as a researcher as an interpretive 'bricoleur' and, hence, the outcome of my writing as a result of my interactions within the field and its participants (Denzin & Lincoln 2011, 4–6). According to Denzin and Lincoln (2011, 4), an interpretive bricoleur produces a bricolage, that is, a 'pieced-together set of representations that are fitted to the specifics of a complex situation'. Similar to how cinematographers edit images and invite 'viewers to construct interpretations that build on one another as a scene unfolds', I have edited the presentation of my results in a certain way (Denzin & Lincoln 2011, 4): one that sometimes tells the same situation from different points of view – the care workers', the care agents', and my own – and one that invites the reader to follow the itineraries care workers take into live-in care work in Switzerland and grasp the infrastructure that enables their migration.

Structure of the book

In order to present how migration for live-in care work is organised and facilitated, along with its broader societal significance, this book is structured in the following way. Chapter 2 presents the social and institutional context in which live-in care labour migration and the home care agencies emerged. From Chapter 3 on, the book presents the empirical findings of my research. Chapter 3 demonstrates how care agents started home care businesses and gave rise to packaged home care services. Subsequently, I explain in Chapter 4 how these agencies are regulated. Moreover, I show the consequences of the interplay between home care businesses and different regulations on the working conditions and migration pattern of live-in care workers. Chapter 5, 6, and 7 present the different steps that care workers undergo on their journey into home care work. Chapter 5 specifically focuses on how care workers learn to search for live-in care work and find employment, while Chapter 6 examines how care agencies recruit care workers by collaborating with recruitment partners and building pools of care workers. I also give insight into how care agents and recruiters select and subjectivise care workers along lines of the gender, age, and recruitment places of prospective care workers. In Chapter 7, I first show how care agents match care workers with care recipients. Subsequently, I trace how care workers travel to their workplace and how care agents organise their travels. Chapter 8 presents the last step in care workers' journeys, illustrating how care workers integrate into the household after arriving at the doorstep of their working place. In Chapter 9, I show the importance of a support infrastructure and the role of care agencies in supporting care workers to create a caring relationship and, hence, a well-functioning care arrangement. Finally, Chapter 10 discusses the results from a migration infrastructural approach, summarising how the migration infrastructure produces a specific form of labour migration, characterised by short-term and just-in-time stints and precarious work. Furthermore, I reflect on the implications of my findings for shifting power relations in labour migration control. The chapter finishes with concluding remarks on future prospects of live-in care labour migration.

2 Paving the way to live-in care work

In the previous chapter, I introduced the global dynamics of gender and socio-economic inequalities that underpin labour migration in the care and domestic work sector. The global rise of migrant domestic workers in the 1990s was mainly attributed to the facilitation of local mothers into labour markets, and discussions focused largely on child care and perceptions of domestic work as 'dirty work' (Constable 1997b; Ehrenreich & Hochschild 2003; Huang & Yeoh 1996; Parrenas 2001; Hondagneu-Sotelo & Avila 1997; Hondagneu-Sotelo 2000; Cox 2006; Pratt 2012; Momsen 2012). However, since the turn of the century, interest in migrant care work specifically for elderly and long-term care has been growing in both research and public discussions (Huang, Thang, & Toyota 2012; Shutes & Chiatti 2012; Lin & Bélanger 2012; Panayiotopoulos 2005; Ayalon, Kaniel, & Rosenberg 2008; Ayalon 2009, 2010; Christensen & Pilling 2018). Considering that the global population is ageing rapidly, this is not surprising. The number of people aged 60 years and above is predicted to grow much faster than the numbers in any other age group (United Nations 2015). These predictions are significant; only 50 years ago, the number of older people was a considerably smaller than any other age cohort. People aged 25–59 outnumbered people aged 60 years or over by 5 to 1, and there were three times as many children younger than 10 as older persons in 1960 (United Nations 2015). By 2030, one-third of the global population is predicted to be aged 60 years and above, and high-income countries are predicted to age the most. Consequently, the provision of care for the elderly has moved to the forefront of policy agendas and public discussions in the Global North. These developments have been accompanied by a number of private market research institutes selling senior health care market forecasts (Freedonia 2017; Global Market Insights 2018; Future Market Insights 2019; Transparency Market Research 2016). The potential of the growing elder care market has been tapped by private recruitment agencies placing migrant care workers into private households specifically for elder care all over the world.

Much has been written on the diverse institutional, structural, and discursive contexts that led to a care gap in private households, even before the topic of ageing societies became as important as today. A surge in the adult worker model in many countries in the Global North (Raghuram 2016) and an increase in the number of women entering the labour market without 'a significant rebalancing of their care costs and responsibilities' (Williams 2011) paved the way for hundreds of thousands of women from lower-waged countries to work as migrant domestic workers. While many women worked and still work in informal employment relations and/or as illegalised residents (Lutz & Palenga-Möllenbeck 2010; Ruhs & Anderson 2006; Anderson 2000; Hess 2005; Metz-Göckel, Münst, & Kałwa 2010), some governments – for example in Canada, the US, Singapore, Taiwan, Hong Kong, Cyprus, countries in the Middle East – have reacted to the care deficit by entering into bilateral agreements with recruitment countries and introducing specific migration schemes for care workers (Cheng 1996; Pratt 2005; Paul 2017; Anderson & Shutes 2014; Silvey 2004; Fernandez & De Regt 2014). 'While it may be considered unusual to consider domestic workers as eldercare-givers in some societies', it is not unusual for private households to employ migrant domestic workers to care for the elderly in these countries (Huang, Yeoh, & Toyota 2012, 196).

Europe has seen a rise in women from both Eastern European countries, working as care workers in the wealthier EU countries, and Latin America, Asia, and Africa, moving to the EU countries (Williams 2011; Lutz 2008). Initially, care workers worked mainly in informal arrangements and found work through informal networks (Krawietz 2014). As early as the 1990s, migrant workers had already established systems for going back and forth between their homes in Poland and their workplaces in Germany (Irek 1998, 75; Morokvasic 1994; Schilliger 2014, 20). The introduction of the free movement of workers in Europe has led to a rise in commercial agencies placing care workers from Eastern European countries in private households in German-speaking countries (see Bachinger 2009, 2010; Krawietz 2010, 2014, 13; Österle, Hasl, & Bauer 2013; Haidinger 2013).

In Switzerland, the emergence of live-in care work and home care agencies was first documented in 2009, mainly in informal care arrangements (Schilliger 2009). However, since 2011, the number of private for-profit agencies placing migrant care workers in Swiss households has mushroomed (Truong, Schwiter, & Berndt 2012). This chapter provides the institutional background that enabled the development of a live-in care market in Switzerland. I first present the context in which home care agencies offering live-in care emerged in Switzerland. Subsequently, I outline key developments in the live-in care market and show how live-in care work is becoming socially acceptable in Switzerland.

A socially constructed care gap

Similar to the developments in other countries in the Global North, recent changes in the Swiss care regime, migration regime, and work regime have paved the way for private households to employ migrant workers. An ageing society, limitations of informal capacities for care in families, and economisation and privatisation trends in elderly care have contributed to an increasing demand in private home care (Greuter & Schilliger 2010; Schilliger 2014). In this sense, the overarching context leading to a care gap does not differ much from other countries in the Global North. Yet, as Huang, Thang, and Toyota (2012) emphasise, care as a concept does not travel in a universal manner. It has to be grasped in its local context, because 'how care is understood and experienced is shaped by social and political–economic contexts operating at the level of the individual or wider society' (Milligan & Wiles 2010, 738). In the following section, I sketch the basic historical and institutional background of the emergence of home care agencies and live-in care in Switzerland. I do so by outlining three areas wherein different logics in developments collide or collude with each other and mirror societal contradictions in the organisation of care and social reproduction.

Feminisation of labour and (neo)liberal practices in the care regime

The first area concerns colliding logics between gender mainstreaming policies intended to promote women in the labour force, on the one hand, and an ageing population, on the other. In other words, while the capacity for unpaid domestic care is decreasing, the number of elderly people in need of care is expected to increase. At the same time, the state has not introduced any significant additional measures to support families or households in providing care. On the contrary, Schwiter, Berndt, and Truong (2018) demonstrate that the Swiss health system has been subject to neoliberal practices and declining public care. This development is explained in more detail in the next sections.

One of the basic characteristics of Switzerland's care policies is that care has always been considered a predominantly private matter (Schilliger 2014, 109; Schwiter, Berndt, & Truong 2018, 5–6). The majority of adults in need of care live in a private household and are mainly cared for privately by their partners (EGB 2010, 13). If partners are not available or able to undertake care, it is mostly the children of the cared-for, usually daughters, that take over this responsibility (EGB 2010, 13). According to a statistical study by the Organisation for Economic Cooperation and Development (OECD) on long-term care in Europe, Swiss households contribute privately to around 60% of care costs (OECD 2011, 47). Thus, the personal contribution of private households to the costs of long-term care is relatively high, especially in contrast to other OECD countries, where long-term care is largely covered by public funding. Even in comparison to other OECD countries with high contributions to long-term care, such as the United States with 40% and Germany with 31%, such contributions are still by far the highest in Switzerland (OECD 2011, 47). This includes the costs that are paid by private health insurance. However, whereas medical treatments are covered by mandatory health insurance, assistance with activities of daily life such as domestic work and help with dressing has to be paid by care recipients themselves (Schilliger 2014; Schwiter, Berndt, & Truong 2018). As a consequence, private households bear a large share of the costs of long-term care out of their own pockets.

In recent decades, the Swiss labour market has experienced a profound structural change in relation to gainful employment. Overall strong growth in part-time employment in families, particularly for women but to a lesser extent also for men, indicates a shift to new forms of family organisation (SFSO 2016, 35). The number of families with a full-time employed man and a non-employed woman primarily responsible for care work, a male earner–female carer model, has decreased since 1970. Instead, the adoption of a one-and-a-half earner model, with a man working full-time and woman working part-time, has increased. Although an egalitarian family model in which both men and women are part-time earners and part-time carers has also become more popular, it still presents only a small proportion of all families (SFSO 2016, 35). These trends reflect the emancipation of women and their inclusion in the labour force in the past few years, which has been high on the political agenda in gender-mainstreaming policies (Lutz 2008, 5; see Schälin 2008 for an overview on gender equality developments in Switzerland). The percentage of women entering the paid workforce has increased dramatically, from 45% in 1970 to 71% in 2016 (SFSO 2016, 33). However, despite the increasing tendency of men and women to share the responsibilities of care work, it is still women that are mainly responsible for its organisation and implementation (SFSO 2014b, 2014a).

With this in mind, Switzerland is no exception in being an ageing society. The generations that were born between World War II and the mid-1960s – so-called baby-boomers – are predicted to retire within the next 30 years (FDF 2016; SOCZ 2008). While birth rates have been low in the last 30 years, life expectancy has risen consistently. The Swiss Federal Statistical Office predicts that the population aged 80 and older will more than double within the next 50 years; more than one million people will be older than 80 by 2060 (SFSO 2010, 28). In addition, more of the elderly are expected to suffer from dementia and related diseases that require extensive care (Höpflinger 2011). According to Schilliger (2014, 99), these developments have been constructed as a 'care crisis' in public discussions. The demographic change is perceived as a problem, even depicted as a threat, with age seen as a burden on the public budget that slows economic growth. She states that the elderly would be treated in a degrading manner, as a handicap burdening the young, and held responsible for an 'explosion' in government spending, health costs, and future pension deficits (Schilliger 2014, 99). In this sense, an ageing population serves as a scapegoat that obscures the dynamics underlying the so-called care crisis.

Instead of expanding the public health system, Schwiter, Berndt, and Truong (2018, 6) find that the state took measures that helped create the conditions for private for-profit home care agencies to flourish. Rollbacks of government involvement in public care provision through neoliberal restructuring processes facilitates migrant care labour, as Misra, Woodring, and Merz (2006) argue in their studies on France and Germany. In the Swiss case, the government introduced new public management reforms with the goal of increasing efficiency in health care, putting hospitals in competition with each other, and reducing health costs. Consequently, hospitals introduced a billing system based on a flat rate per case and adhered to limited budgets (Schwiter, Berndt, & Truong 2018). Instead of calculating the cost of effective treatments for an individual patient, hospitals now receive a fixed amount of money depending on the diagnosis. This arrangement is supposed to induce a faster hospital discharge after surgery or other treatments to reduce the number of nights that patients spend in hospital (Schilliger 2014, 113). The shorter and the simpler the treatment, the more profit a hospital can make. As a result, as Schwiter, Berndt, and Truong (2018) state, part of the burden of rehabilitation has been shifted from public health institutions to private responsibilities

This additional care is assumed to be performed by family members or acquaintances (Greuter & Schilliger 2010). If capacity for this is lacking, home care services, such as outpatient care agencies offering hourly care, may be used. However, outpatient care services have been subject to rationalisation processes similar to hospitals and other public health institutions, such as nursing homes. When outpatient care services were included in the coverage of health insurance in 1996, the providers had to introduce a new system of allocating costs. This new system distinguishes medical treatments that are covered by health insurance from other care services that have to be paid by the care recipients themselves (Schilliger 2014, 125; Schwiter, Berndt, & Truong 2018, 6). The providers of outpatient care services had to adopt a rigid time control scheme, in which care workers adhere to a prefixed duration for a specific task (Greuter 2010, 106). As a consequence, outpatient care workers have less time to spend on care work in individual households. Home care agencies selling live-in care, on the other hand, market exactly the opposite: time for care (Schilliger 2014, 11). Hence, care agencies arrive on a scene in which diverse labour market and health care restructuring has contributed to the construction of a care supply gap. Home care agencies promise to close this gap by offering live-in care services as individual solutions.

However, it is important to keep in mind that the care crisis is not merely an outcome of specific austerity programmes or neoliberal practices in welfare states, but that it is inextricably linked to larger socio-economic developments that sustain gendered, racialised, and socio-economic inequalities in care provision. As Fraser (2016, 117) states, 'the roots of today's "crisis of care" lie in capitalism's inherent social contradiction'. More specifically, she argues that it is deeply rooted in the social order of contemporary neoliberal and financialised capitalism, which subjugates social reproduction to the realm of production (see Fraser 2014a, 2014b, 2016).

Gendered division of labour and limited labour protection in the household

The second area concerns the societal understanding of work in a private household, which plays into home care agencies' marketing strategies. In their famous essay, Bock and Duden (1976) demonstrate and criticise how domestic work was transformed into a 'labour of love' and constructed as a natural domain of women in the course of the capitalist development of industrial societies. Subsequently, domestic work was politicised in campaigns calling for waged domestic work and discussed in terms of the relation between production and social reproduction (Cox & Federici 1976; for an overview see Kofman & Raghuram 2015; Schilliger 2014). Since then, feminist scholars have challenged the construction of a duality between women's undervalued and unpaid work, ascribed to a private sphere, and men's wage labour, ascribed to the public sphere, as well as the implicit underlying gender contract, in many ways (Lutz 2008, 2012; McDowell 2009; McDowell et al. 2011). According to McDowell et al. (2011, 220), this gendered division of labour in the labour market and in domestic work has been part of the 'prevailing ideology and state policy in many western societies'. Moreover, because of its association with the supposedly 'natural' attributes of femininity, she stresses that care work is undervalued when the exchange takes place in the marketplace (McDowell 2009, 82). As a consequence, wages are usually low for such exhausting and difficult working conditions. Care agencies capitalise on this language of care as a labour of love by marketing 'worker's love' on their websites (Pelzelmayer 2017, 12).

Another key aspect of care work in private households is that working conditions in many countries are characterised by a low level of labour regulation and relatively low salaries in comparison to other sectors. Consequently, as Chen (2011, 170) observes, live-in domestic and care workers face greater isolation, limited mobility, long working hours, and greater exposure to abuse. Truong (2011) and Schilliger (2014) have demonstrated that the live-in situations of care workers in Switzerland blur the boundaries between work and private life, which makes it difficult to distinguish working hours from free time. Huang and Yeoh (2007, 212) show that abuse of domestic workers in Singapore can take place 'under the discourse of family' and is perpetrated by women employers on a basis of class asymmetries, ethnicity, and nationality. Moreover, they stress that uneven power relations in the household as a workplace and the characteristics of domestic work can hinder domestic workers' desires to leave their employment and to claim social justice in cases of abuse (Huang & Yeoh 2007, 212; Yeoh, Huang, & Devasahayam 2004, 16). In contexts where domestic workers' only opportunities for employment are through recruitment agencies and debt bondage, they are also more exposed to abuse from employers (Chen 2011, 170). Examples of such abuse include the withholding of wages and passports.

In Switzerland, in contrast to other service work and the industrial sector, care work in direct employment in private households is not subject to a generally binding collective agreement or to labour law. The only other sector in Switzerland that is exempted from a collective agreement and labour law is agriculture (see Chau, Kumar, & Lieberherr 2015, 3). According to Schilliger (2014, 147), this exemption is a relic of the past. More specifically, it is the legacy of a time when the employment of maidservants was common in agricultural and in bourgeois households and was regulated through cantonal provisions for servants (Rippmann 2011; for an analysis of German domestic servants in historical perspective, see Lutz 2012). Today, the working conditions in households are regulated by standard employment contracts (*Normalarbeitsverträge*) that nonetheless differ in each canton. In contrast to collective agreements regulated by labour law, standard employment contracts allow deviations at the expense of the workers. Therefore, they provide weak protection for workers, and claims to labour-related rights result in individual case suits in which the burden of proof falls on the plaintiff employees (Medici 2012, 8; Schilliger 2014, 147–50). By minimising

employers' legal risks and thus potential legal costs, the low protection of employees in a household and the live-in situation of care workers enable care agencies
to market 'affordable care', in other words, low-priced care, on their websites
(Pelzelmayer 2017, 12).

Utilitarian migration regime and home care business development

The third area refers to the compatibility between migration regulations and the
development of commercial home care agencies in Switzerland. The form of
mobility that live-in carers experience, going back and forth between place of
residence and workplace across national boundaries, is fostered by the agreement
between Switzerland and the European Union on the free movement of workers,
which was introduced in 2002. It allows EU and European Free Trade Association (EFTA) members to live and work in Switzerland. In May 2011, freedom
of movement was extended to eight new EU member states. Since then, citizens
from Estonia, Latvia, Lithuania, Poland, Slovakia, the Czech Republic, and Hungary have been granted facilitated access to the Swiss labour market. This relaxation of admission regulations was crucial to the emergence of home care agencies.
In order to understand this relaxation and its meaning for the commercialisation of
live-in care work, however, it is not enough to simply examine the recent development towards freedom of movement; we need to look at the historical development of Swiss migration regulations.

In the first decades of the nineteenth century, Switzerland had no centralised
immigration policy. After the establishment of the Swiss Federation, liberal
migration treaties were concluded in which natives and foreigners would be economically and legally equal in every respect (Hoffmann-Nowotny 1985, 207).
Originally designed to ensure legal emigration from Switzerland, these treaties
also provided a framework for the immigration that began in the second half of
the nineteenth century. The rapidly increasing number of migrants induced the
expression *Überfremdung*, a term that literally translates as 'overforeignisation',
in public discussions, which have accompanied changes in Swiss migration policy
intermittently from then on (Hoffmann-Nowotny 1985, 206).

At the end of World War II, Switzerland was confronted with an acute short-
age of labour. The government reacted by pursuing large-scale labour recruitment by means of bilateral agreements with countries such as Italy and Spain
(Piguet 2004, 16). These agreements arranged the short-term immigration of
'guest workers', mostly men, who often left their families behind to work in
construction, agriculture, and industry (Piguet 2006, 70). As the French term for
them, *saisonniers* (seasonal workers), implies, they could not stay all year round,
and they had to renew their permits every year (ANAG 1931; Arlettaz 2012).
As well as permanent settlement, these workers were forbidden from changing
employment or bringing their families with them. The post-war economic boom
was expected to last only temporarily, so the immigration policy aimed to 'rotate'
workers to prevent them from staying permanently (Hoffmann-Nowotny 1985,
216). In periods of high demand for labour, D'Amato (2008, 37–38) argues, the

borders to Switzerland were opened, while in downturn periods the borders were closed.

From the beginning of the 1960s, the liberal admission policy became the subject of growing criticism (Piguet 2006, 71). According to Piguet (2006, 71), rising inflation was discussed as a consequence of the additional demand by foreign workers for goods and services, and as a result, expressions of xenophobia became more frequent. To counter initiatives of anti-foreigner movements demanding a massive reduction of the foreign population, the government introduced a global ceiling for migration and a central register of foreigners (*zentrales Ausländerregister*) (Piguet 2006; Wicker 2010). Cantons, communities, and federal offices were required to regularly provide information about the national origin and legal status of foreigners. Therefore, the instigation of a global ceiling established a control instrument for the Swiss nation state to monitor its foreign population.

As a consequence of the depression in the 1970s, the number of migrants decreased for the first time since World War II (D'Amato 2008). In this sense, D'Amato (2008, 38) argues, the function of migrant workers as an 'economic cycle buffer' has never been more obvious than it was after the first oil shock in 1973. Through the non-renewal of the permits of unemployed foreigners, the state had the opportunity to reduce its labour force without increasing domestic unemployment (Piguet 2006). During the 1980s, as the economy recovered, the government was again interested in meeting the growing labour demand of the economy, so the global ceiling was officially maintained. Between 1985 and 1995, about 40,000 new labour permits were issued yearly, and 130,000 seasonal workers were legally allowed to enter Switzerland (D'Amato 2008). Since Italy and Spain had ebbed as recruiting sources, a large proportion of the new immigration came from Yugoslavia and Portugal (Piguet 2006). Similar to past periods, the migrant labour force undertook mostly less qualified occupations in migration-dependent sectors.

During the 1990s, various developments induced intense political friction regarding migration policy. One development was a relative long period of economic stagnation, whereby unemployment increased because migrant workers did not act as 'cyclical shock absorber[s]' anymore (D'Amato 2008, 38). In fact, they gained social rights and improved residence permit status in the meanwhile, as the countries they emigrated from negotiated improved residence conditions for their citizens in Switzerland (Stephen & Miller 2009; Piguet 2004). Moreover, humanitarian considerations and international pressure forced Switzerland to deal with migration forms that did not relate directly to the labour market, and the influence of international right and international treaties had to be included in the planning of migration policy (D'Amato 2008). The topic of refugees and asylum seekers became the main issue of migration discussion in the 1990s, and in consequence, immigration policy became subject to increased politicisation.

In reaction to these developments, the government worked out reform proposals for migration policy and introduced a 'dual circle model'. The model differentiates migrants by country groups. While nationals from the EU and EFTA countries benefit from free movement with Switzerland, nationals from all other states should not be able to migrate to Switzerland, except in very specific cases of highly skilled labour or as refugees seeking asylum. The dual system is based on the idea that workers from within the EU can fulfil a labour demand for low-skilled workers, as it was assumed that European workers and highly qualified workers would arouse less resentment in the population (D'Amato 2008, 41). The development of the freedom of movement is closely related to the idea of high 'labour mobility' of EU nationals within the European Union and EFTA states (Bauloz 2016, 5–9). The idea is that EU nationals are free to move between their places of residence and workplaces across European countries. A number of researchers have observed that the circulation of migrant workers, not just in Europe but also in other parts of the world, has increasingly been promoted as a migration policy tool to support development in home countries, meet labour market needs without permanent settlement, and minimise irregular migration (Triandafyllidou 2010; Vertovec 2007; Wickramasekara 2011). In migration and development discourses, it is claimed that the circulation of migrant workers generate triple-win situations (Castles 2006; European Migration Network 2011; GCIM 2005; Wickramasekara 2011). This win–win–win logic of circular movement has also been adopted by individual private home care agencies:

> For many Swiss, it is not possible to afford 24-hour care from local service providers for financial reasons. (. . .) The care workers from GETcare are in a similarly difficult situation. Care employment offers them the opportunity to work in the occupation they were trained in and additionally earn a much higher salary than in their home country. Hence, after their stay in Switzerland – where they can get to know a new culture besides work and improve language skills – they go home to their home country with a good financial situation. Therefore, both parties benefit.
>
> (website of a care agency, getcare.ch, 2015;
> translated from German by author)

My main point here is to show that Switzerland has developed a utilitarian migration policy that has played an important role in the emergence of home care agencies and their business practices, as I will show in later chapters. The historical overview of Swiss migration policy demonstrates that two elements were the main driving force of Swiss migration policy. On the one hand, grassroots movements and xenophobia tried, through direct democracy, to block immigration by putting a maximum ceiling on foreign population to the vote. Such attempts have been made nine times during the last 50 years (EKM 2015; Piguet 2006). The most recent, launched by the right-wing Swiss People's Party [*Schweizerische Volkspartei*], was accepted by popular vote in 2014. It set politicians the difficult task of reconciling how to limit the number of EU nationals in the Swiss workforce without breaching the free movement of persons treaty that Switzerland has signed with the EU (Schweizerische Bundeskanzlei 2017). This would have led to a termination of all other EU agreements and hence risked trading partnerships. After two years of negotiation, the Swiss government decided to reject quotas and introduce a series of soft measures targeted at privileging unemployed locals (Bundesrat 2019). However, dissatisfied with this compromise, the Swiss People's Party has already launched another referendum to limit migration, this time with the explicit demand that Switzerland forfeit its relationship with the EU (Bundesrat 2018). Thus, the Swiss political system of direct democracy gives migration policy particular salience in public debate and constantly puts the government under pressure. On the other hand, the Swiss government has historically opened and closed its borders according to its assessment of the economic demand for labour and mainly used utilitarian arguments to pacify voices in the population speaking out against labour migration. Although the latest development of the freedom of movement attempts to harmonise social rights and labour market access within the European Union and therefore de jure creates more equal working conditions, it follows precisely this utilitarian logic and aligns with home care agencies' offer of supposedly affordable 24-hour care by promoting the rotation of migrant workers.

It is in this context that care agencies offering live-in care have emerged. It is a context in which family members have largely performed elderly care in the past and government support is considered a last resort for those without the financial means or social resources to organise private care. As the amount of care for the elderly is expected to rise and responsibility for care is increasingly being shifted from public institutions to private households, home care agencies offer individual solutions to fill the socially constructed care gap. Moreover, the low level of labour regulations applying to private households plays a significant role in helping home care businesses offer relatively low-cost live-in care packages and market home care on an understanding of care as a 'labour of love' (Bock & Duden 1976). Finally, the emergence of home care agencies is facilitated by a utilitarian migration policy that promotes high labour mobility between countries in the freedom of movement framework to enable economic development.

How live-in care work is becoming socially acceptable

The development of private for-profit live-in care businesses has not gone unnoticed. Indeed, it has moved to the forefront of mass media and public discussions. A documentary film on live-in care workers (Batthyany 2013), a public discussion between experts in a TV debate show (SRF 2013c; Rundschau 2011), and countless contributions in newspapers and on TV news about live-in care (among others SRF 2013a, 2013b) all indicate that new social changes are taking place. Schwiter, Pelzelmayer, and Thurnherr (2018) argue that the dominant media discourse of a booming live-in care market also produces a self-fulfilling prophecy and contributes to the diffusion of live-in care. Indeed, as a care agent I interviewed, Mattea, explained when I asked her about her negative portrayal on TV as one not complying with Swiss regulations, said, 'there is no negative publicity, only publicity'.

Moreover, several organisations have shown interest in live-in care work. The Department of Equality of the City of Zurich commissioned a study, which was carried out by the Geography Department at the University of Zurich and published in 2012 (Truong, Schwiter, & Berndt 2012). Subsequently, the Department of Equality launched a website called CareInfo.ch, which provides information for care recipients, care workers, and all other interested parties. It is currently in the process of expanding collaboration with departments in other cities and cantons, as the project leader told me in an informal conversation (see CareInfo 2013).

Furthermore, trade unions such as VPOD and UNIA took on home care work as a new sector to represent and started to organise care work. VPOD, in collaboration with care workers, most prominently Bozena Domanska, formed a group called Respekt (Respekt 2015). The group, initially mainly live-in care workers from Poland, meets once a month after church on Sundays to discuss their work situations and how to improve them, exchange information, and support each other in finding jobs. It also provides advice and legal consultancy for individual live-in care workers. The group was initially founded after Bozena Domanska filed a lawsuit against her former employer for wrongful termination by a care agency. Bozena Domanska and Respekt attracted substantial attention from the media. She was portrayed in a documentary film (Batthyany 2013) and later had numerous appearances in TV news, shows, and newspapers. Respekt advocates not just for improved working conditions but also to achieve more recognition for care work in society generally (for an extensive overview of the development of Respekt, see Schilliger 2015; see also Chau, Pelzelmayer, & Schwiter 2018).

The working conditions and legal protection of live-in care workers have slightly improved following the adoption of various measures in response to the emergence of home care agencies. As a supplement to the free movement of persons in 2002, the government introduced measures designed to prevent wage dumping in 2004 (EntsG 1999; EntsV 2003; FDFA 2019; SECO 2017). The goals of these accompanying measures are to protect national salaries and working conditions and to secure equal conditions of competition for domestic and foreign businesses (SECO 2017). In response to a study reporting repeated undercutting of customary salaries in domestic work (Flückiger 2008), the Federal Council introduced a national standard employment contract specifically for private households in 2011, initially until the end of 2013. It stipulated minimal hourly wages for live-in care workers, with a first level of CHF 18.20 for unskilled workers, CHF 20 for unskilled but experienced workers, and CHF 22.00 for workers with a Swiss Federal Certificate in domestic care [*Hauswirtschaft*] (NAV Hauswirtschaft 2010a). In 2013, the Swiss Federal Council decided to extend the national standard employment contract until the end of 2016 and raised the minimum salary by 35 centimes (Bundesrat 2013; NAV Hauswirtschaft 2010b). In 2017, it was again extended for another three years (Bundesrat 2017b).

Care workers that are employed not by the households directly but rather by placement agencies are subject to a collective agreement that came into force in 2012 for the entire field of work placement (GAV 2012; Bundesrat 2011). At a minimum of CHF 16.46 per hour, its salaries are lower than the minimum in the standard employment contract. In its stipulation of maximum working hours, overtime, night work bonus, and Sunday bonuses, it is de jure an improvement for live-in care workers employed with an agency. However, de facto, it is difficult for care workers to claim labour-related rights, and care workers often work much more than the hours stipulated in their contracts (see Schilliger 2014, 152–53).

So far, there has been one case in Switzerland where a care worker, with support from Respekt, brought her case to the civil court of Basel to sue a placement agency for working hours and presence time without being paid. The care worker was able to prove her additional working hours, and the court ruled in her favour. The agency had to compensate additional working hours in full and pay half of the usual hourly wages for every hour spent on call. For three months of live-in care work, the agency had to pay an additional amount of CHF 17,000 (Schilliger 2015). Respekt has now employed a lawyer to take over all the complaints being filed against agencies. Most of the disputes are resolved by settlement before going to court, as the lawyer told me in an informal conversation in May 2017. It seems likely that there are many more disputes than we know of between care workers and care agencies that are resolved by settlement.

Meanwhile, the UNIA union has negotiated a collective bargaining agreement for German-speaking Switzerland with the employers' association Zu Hause Leben. This association at the time consisted of approximately 25 care agencies, most of which were branches of one large care agency (UNIA 2014; UNIA & Zu Hause Leben 2014). The agreement stipulates a minimal salary of CHF 22 per hour and a full thirteenth monthly salary (UNIA & Zu Hause Leben 2014). However, as one of the unionists told us during an interview, in the end the agreement was not adopted because too few care agencies supported it.

Another development worth mentioning is the meeting at a national conference in May 2017 of a range of organisations, from drop-in centres for undocumented migrants, known in Switzerland as *sans-papiers*; to departments of equality of various cantons and cities; organisations supporting migrants, such as the Women's Information Centre (FIZ); unions; and researchers. It was organised by the Department of Geography at the University of Zurich and a left think-tank called Denknetz. The goal of the meeting was to network and brainstorm further activities to improve the conditions of paid domestic and care work in private households. It also provided a platform to connect the topic of domestic work, which is often provided by *sans-papiers* in Switzerland (Alleva & Niklaus 2004; Flückiger & Pasche 2005; Flückiger 2008; IGA 2007), and live-in care work, known to be provided more often by cyclically migrating workers. Hence, new alliances emerged between *sans-papiers* organisations and organisations for live-in care work, which until then had often been treated as two separate phenomena. This could be because the topic of *sans-papiers* domestic workers seems to be publicly debated more often in French-speaking western Switzerland, whereas live-in care work is more frequently discussed publicly in German-speaking Switzerland (Schwiter, Pelzelmayer, & Thurnherr, forthcoming, 8). Therefore, for these women's and migrant organisations, unions, and other actors, live-in care work performed by migrant care workers has become a vantage point to take the struggle of gender equality and negotiations over paid and unpaid work to a new level. The Geography Department at the University has also been working on a follow-up research project called Decent Care Work? Transnational Home Care Arrangements, in collaboration with Helma Lutz at the University of Frankfurt and Brigitte Aulenbacher at the University of Linz, on expectations of good care and good work.

Finally, a member of the National Council, Barbara Schmid-Federer, filed a parliamentary request to the Federal Council seeking to improve live-in care workers' working conditions in 2012 (Schmid-Federer 2012). Subsequently, the Federal Council issued a report on the legal framework for circular migrant care workers in 2015 with five possible regulations: subordination of live-in care employment to labour law, creation of a new decree regarding labour law, modifications to standard employment contracts, the creation of a collective agreement, and the introduction of a duty on employers to provide information about working conditions (SECO 2012). Consequently, the Federal Council commissioned a research and consulting agency to assess the impact costs of these regulation scenarios. The subsequent report presented three scenarios, in which either 10%, 50%, or 100% of on-call time was calculated and remunerated as working time (B,S,S. 2016). It concluded that the scenario with minimal regulation (10%) would lead to additional costs of up to CHF 15 million and the scenario with maximum regulation (100%) would lead to CHF 30 to 90 million in additional costs for the public sector and health insurance companies (B,S,S. 2016). Moreover, the report starts from the assumption that higher costs would lead to an increase of informal employment, especially if the middle or maximum scenarios were to be applied. However, the reasons for this assumption are unexplained. Based on this report, in June 2017 the Federal Council decided to introduce the minimal regulations suggested by modifying the cantonal standard agreements by mid-2018 (Bundesrat 2017a).

In sum, the emergence of home care businesses brokering migrant live-in care workers has triggered a range of new developments in the social and regulatory dimensions of live-in care work. The negotiations over working conditions encouraged the formalisation of live-in care employment of migrant workers in Switzerland and, hence, a legitimation of the live-in care market and commodification of care. Overall, current developments are likely to result in general social acceptance of the live-in care model in Switzerland. While the majority could not even have considered living with and being taken care of by a migrant worker only a decade ago, migrant care work is now slowly developing – though not without friction and struggles in academic and public debates, as demonstrated earlier – into a socially and broadly accepted solution for elderly care.

3 The rise of home care agencies and packaged home care services

Although the topic of live-in care work has only recently become prominent in public discussions and the media, it is important to keep in mind that live-in care work in Switzerland is not entirely new. It has existed for quite some time through informal arrangements. Many care workers told me stories about informal working arrangements, some of which dated to the early 2000s. A care agent also told me that he had employed a live-in care worker from Slovakia to take care of his mother-in-law for eight years before he started his own care agency in 2008. That was long before the regulations allowing free movement of workers came into force in Switzerland in 2011.

What is new, however, is the availability of package deals offered by a growing number of home care agencies specialising in non-medical care and live-in arrangements. Designed as all-inclusive offers, the packages include the recruitment and matching of a care worker, the organisation of their travels, and the option of replacing care workers in case their relationships with care recipients do not develop as expected. What is also new is the speed with which such care arrangements can be organised. 'Uncomplicated and fast', one of the agencies advertises, 'it takes a maximum of 60 minutes of time. You can welcome a carer at your door after 6–10 days'. Some of the agencies have organised care arrangements in as few as three days.

With the development of a broad landscape of agencies, it has become easier for older people in need of care and their families to access live-in care. Information about live-in care is generally easy to access. The agencies post advertisements in newspapers and distribute information brochures at medical practices. Some of the agents I interviewed collaborate with outpatient care services, individual medical practitioners, nurses, and organisations that represent the interests of the elderly (for example Pro-Senectute, the largest expert and service organisation for the elderly in Switzerland). Hence, it is not difficult to make contact with home care agencies in everyday life. More importantly, agencies have provided an abundance of live-in care websites online. Not only is it possible to gather information and compare offers, but their services can also be booked online or over the phone. This chapter provides an overview of the range of home care agencies and services.

Offering all-inclusive deals

'I think Swiss people are always happy when they know straight away how much something costs, without any hidden extras like transportation, insurance, and so on', Dominik said while handing me an information sheet about his services. He was the owner of a small employment agency that places care workers from Slovakia. His agency offers three care packages that provide differing levels of care dependency (Table 3.1).

'Everything's included: VAT [value added tax], taxi rides, the care workers' pay, all social benefits, occupational accident insurance, non-occupational accident insurance, and sick pay', Dominik added. Other care agencies offer up to five different price models, depending on 'what experience, German-language skills, and education the client wants the care worker to have', Daniel, owner of an employment agency that places care workers from Poland, explained to me. The care packages usually encompass finding and matching care workers to households, administrative work such as wage accounting, registration with social security offices, taxes and applications for residence permits, transportation to care recipients' households and back home, and their replacement, if needed. Some more basic packages consist of the placement and the organisation of the care workers' journey. Some care agencies provide administrative support on top of these basic offers. For example, Dora, who runs a one-woman employment agency, provides advice on the application for work permits, salary administration, and the registration with social insurance upon request. 'Most of the time, they [the care recipients and family members] don't want to do it [administrative work] themselves', she explained. At the time of the interview in 2014, she charged CHF 45 per hour for this service. 'I've done several registrations, so I'm more familiar with this now than people who've never heard of it and don't know which forms to use. People are happy to pay for it; it's much easier for them', she said.

Table 3.1 Price information sheet of a care agency, November 2014.

CHF 205 per day **(around CHF 6,150 per month)**	Level 1: light care People who need assistance with personal hygiene, nutrition, or mobility at least once a day and also need assistance in domestic care several times a week.
CHF 225 per day **(around CHF 6,750 per month)**	Level 2: intermediate care People who need assistance with personal hygiene, nutrition, or mobility at least three times a day at different times of the day and also need assistance in domestic care several times per week.
CHF 245 per day **(around CHF 7,350 per month)**	Level 3: intensive care People who need assistance with personal hygiene, nutrition, or mobility around the clock, including at night, and also need assistance in domestic care several times a week.

Source: (Translated from German by author.)

Prices and the scope of services

The prices for live-in home care start at around CHF 1,600 and can rise to CHF 25,000 per month. While the lowest offers, sometimes referred to as cheap care (*Billigbetreuung*) and cheap labour (*Billig-Arbeitskräfte*) by the media (Bracher 2011; Kislig 2015; SRF 2013a; Wehrli 2013), cost between CHF 1,600 and CHF 4,000, most care agencies offer live-in care services in the middle range with costs around CHF 5,000–6,000 per month. One of the largest care agencies in my sample offered packages for around CHF 12,000 per month. However, the most expensive costs for home care that I came across during my fieldwork were between CHF 15,000 and CHF 25,000 per month. This range was quoted by a placement agency located in a prosperous region in Switzerland that specifically targets wealthy care recipients.

According to Dominik's information sheet, the scope of service provided by live-in care workers comprises the following (Table 3.2).

As Table 3.2 shows, the work of the carers consists predominantly of assistance in everyday life in and around a household (see also Schilliger 2014; Van Holten, Jähnke, & Bischofberger 2013). Dominik's information leaflet specifically states that medical care is excluded from the scope of services, for legal reasons. Medical care is regulated in Switzerland and can be distinguished between basic medical care and more advanced medical care. Basic medical is defined as performing tasks in relation to personal hygiene, diet, and movement, if the care recipients are not able to perform the tasks themselves. It is not usually officially part of the live-in care arrangement, as only trained care workers are qualified to provide basic medical care and only certified nurses are allowed to provide more advanced medical care (Medici 2012). To obtain certification for basic medical care, care workers have to complete a course offered by the Swiss Red Cross, comprising a theoretical part of 120 hours and a practical part of 12 to 15 days that costs more than CHF 2,000. Care recipients in need of medical care typically complement their live-in care arrangements with hourly outpatient care performed by nurses. Home care agencies offering live-in care arrangements are not allowed to offer

Table 3.2 Scope of service of a care agency, November 2014.

Housekeeping	grocery shopping, grocery preparation, creating an orderly domestic environment, cleaning, care of clothes, care of plants and pets
Keeping company	being present, eating meals together, organising how to spend time, escorting to doctors' appointments and visits
Providing a helping hand with personal hygiene	assisting with washing and hairdressing, dressing and undressing, helping with physical activities (getting up, going to bed, going to the toilet, preparing walking aid)
Creating security	providing orientation in space and time, calling attention to prepared medication or to reported non-intake of medication, seeking help in emergency situations

Source: (Translated from German by author.)

medical care (*Pflege*) unless they offer medical care services with certified staff. Given that outpatient care services, commonly known as *Spitex* (short for *Spital externe Hilfe und Pflege*), offering both hourly medical and non-medical care are well-established institutions in Switzerland, the majority of the agencies offering live-in care specialise in non-medical care, or in other words, assistance in everyday life (*Betreuung*). Hence, the work performed by live-in care workers in Switzerland is similar to the domestic work described in classic and contemporary literature on global domestic work.

The question remains whether the difference between assistance in everyday life (*Betreuung*) and medical care can be distinguished as clearly in practice. Given the fine line between basic medical care and assistance in everyday life, it is hard to imagine that care workers always know how to react in grey areas, especially if they are provided with little training. Putting medication in front of a care recipient, for example, is allowed for non-certified live-in care workers providing assistance in everyday life. Putting medication into a care recipient's mouth, however, would be considered basic medical care. The differentiation between medical and non-medical care is also interesting, as it reflects how society values the work of trained and untrained care workers. The hourly costs of around CHF 50 for basic medical care are much higher than the hourly costs of around CHF 35–40 for tasks more commonly associated with domestic work. Both costs are quoted for outpatient care services (*Spitex*). Home care agencies such as Dominik's provide affordable solutions for care recipients in need of more assistance than just a couple hours a day.

Salaries and placement fees

What is striking is that, while the care agents freely told me the costs of their services and the sums that the care recipients have to pay, their answers when I asked about salaries mostly remained vague. The only clear answer that I received was from Pascal, owner of an employment agency that recruits from Hungary. He even gave me an anonymised example of a pay slip (Table 3.3).

As can be seen in Table 3.3, care workers with Pascal's agency receive a monthly net salary of CHF 1,736 or EUR 1,400 (rate 1.24). This is much higher than care workers' earnings in the recruitment countries in Eastern Europe, where median salaries ranged between EUR 400 and EUR 800 net in 2014 (Eurostat 2019). However, given that the median salary in Switzerland was around EUR 4,763 gross for women (CHF 5,907) and EUR 5,444 gross for men (CHF 6,751) in 2014 according to Swiss Statistics, or around EUR 3,548 net (CHF 4,400) according to European Union Statistics, this amount is much lower than what locals earn and far from enough to live on in Switzerland (BFS 2017; see Chau, Pelzelmayer, & Schwiter 2018). The lowest net monthly salaries I heard of when interviewing care workers with agencies were around CHF 1,400, which equals what students commonly earn as interns in non-profit organisations during their studies or right after graduation. The highest salary was reported by a care worker earning around CHF 3,300 net, which is the equivalent of a monthly salary at the lower end of the

Table 3.3 Outlays for care recipients provided by a care agency, June 2014.

Costs of domestic help for client		
Pay slip example (exchange rate: CHF 1.24 = EUR 1.00) Name: Anonymised Number of days worked: 20 Employer: Anonymised	Time period: 20.05.13–17.06.13 Number of days for board and lodging: 30	
	Salary received by care worker	**Total cost paid by care recipient**
Gross salary	CHF 3,071.55	
Old age insurance, disability insurance benefits, compensation for loss of earnings 5.15% of gross salary deducted from care worker's salary and another 5.15% to be paid by employer	− CHF 158.18	CHF 316.36
Unemployment insurance 1.10% of gross salary deducted from care worker's salary and another 1.10% to be paid by employer	− CHF 33.79	+ CHF 67.58
Withholding tax 5.00% of gross salary deducted from care workers' salary	− CHF 153.58	+ CHF 153.58
Board and lodging Deducted from care workers' salary	− CHF 990.00	
Total deductions for care worker	− CHF 1,335.55	
Net wage paid to care worker (in Euros)	**CHF 1,736.00** (EUR 1,400.00)	+ CHF 1,736.00
Old age insurance Administrative costs		+ CHF 80
Health insurance Hungary		+ CHF 60
Accident insurance		+ CHF 90
Placement fee		+ CHF 400
Administration and travel costs		+ CHF 1100
Total costs per month for employer		**CHF 4,003.52**

Source: (Translated from German by author.)

service industry. A further point that needs to be taken into consideration is that care workers work on a rotational basis; if they work for one month and stay at home the next one, they are only paid every second month. Therefore, live-in care ranks among the lowest paying jobs in Switzerland.

Some of the care workers experienced higher wages over time and when changing employers. Berta, a care worker from Hungary, received a monthly salary of EUR 1,000 for a temporary employment of three months on her first stint with an agency. When she changed employment to another household with the help of private contacts – hence, leaving the service of the care agency – she received a monthly salary of EUR 1,650. This does not mean that care workers always earn more in direct employment than when working through agencies. Moreover, the value of their salary also depends on their working hours and working arrangements. For instance, the care workers I talked to with one of the larger care agencies worked in two-week shifts; they would work in the households for two weeks and then go back home for another two weeks. With a salary of around EUR 1,600 every month, they earn more for the hours they work than some other care workers who work a month full time. Thus, the wages and working conditions differ considerably within the industry.

Placement fees are usually paid by care recipients. Only one of the businesses in this sample charged care workers a recruitment fee, and that was only upon successful placement. The owner of this agency, Dora, charges the care recipient's family a one-time fee of CHF 4,000 for mediating two care workers and takes 5% of the yearly gross salary from the care workers, which is the maximum fee that job seekers can be charged (AVG 1991, Art. 3). All the other care agencies interviewed only charge the care recipients, not the care workers. However, the margins did not become transparent to me, as not all agents would speak openly about how much profit their services make. My findings indicate that the margins differ considerably. For example, Anthony, the owner of an employment agency, charged CHF 400 for a placement, of which CHF 200 were forwarded to his recruitment partner in Slovakia. With around ten successful placements in total since he had established his agency, his was one of the smaller businesses in the sample. Mattea, owner of a care agency that recruits from Slovakia, charged a CHF 45 initial placement fee and CHF 350 administrative costs every month. In contrast, Pascal, as shown by the pay slip example, charged CHF 400 placement fees from the care recipients and CHF 1,100 for administrative costs and travelling fees every month. Since his own agency operates the transport of the workers he places, it is not very clear how much profit he makes. Hence, there is a diverse range of placement costs, which are not transparent to the public.

While care agencies based in Switzerland generally do not charge care workers a placement fee, care workers do report having paid recruitment fees, albeit with agencies in the recruitment countries and in informal networks. For example, Krisztina, a care worker from Hungary, paid a placement fee of EUR 300 to Interland, a transportation agency in Hungary that is well known by the care workers I talked to for placing care workers in households in Germany, Switzerland, Austria, and Liechtenstein. They are officially registered as a transport firm in Hungary and are hence not allowed by the Swiss authorities to place care workers in Switzerland. Krisztina estimates that Interland places around 300 women from Hungary per year. During the course of my fieldwork, I came across their name intermittently, and I tried to contact Interland. Unfortunately, they did not respond to my overtures. It seems that Interland charges care workers a fee for every placement they make. Krisztina told me about a friend of hers who was placed three times in new households by Interland within one year and was charged EUR 300 every time. She also knows of private intermediaries who place in Germany and who charge the care workers around EUR 50 every month. Care workers also charge each other for information about employment opportunities, as Krisztina explains: 'I know someone, she's been working as a care worker for years. If she hears of a job, she'll ask 200 Euros for the information, just for the phone number. Women do this with each other'. Therefore, although it is not the norm in the Swiss recruitment and placement industry to charge employees a fee, it seems to be more common to do so in the recruitment countries. However, these fees are only a fraction of care workers' monthly salaries and hence relatively low in comparison to the fees of domestic workers based on debt-migration in Southeast Asia and Middle Eastern countries, where agencies commonly ask for recruitment fees of up to several months of salaries (Goh, Wee, & Yeoh 2016; Killias 2009; Parrenas & Silvey 2016).

Why not? An employer's experience with packaged home care

How do care recipients and their family members come to use the service of a care agency to arrange live-in care? This is the story of Mr Schmid, who I found through an online forum on Alzheimer's, and of how he first came to use the services of a home care agency. I visited Mr Schmid on a sunny afternoon in February 2015. He was happy to talk about his experience of live-in home care with me over tea and biscuits. His wife had been diagnosed with Alzheimer's in 2006. 'It'd gradually gotten worse. At some point, it was too much for me. So, I started to do a bit of research online. I just typed in "care workers from the East" (*Pflegerinnen aus dem Osten*)', Mr Schmid explained. Subsequently, he contacted two agencies that he had found. The first one is located in Hungary, the second in Poland.

> Then, I talked to this gentleman [from the agency in Hungary]. He spoke German very well. Of course, that made a good impression too. I had an offer from him and one from the Polish organisation. With them [the Polish one], I would have had to pay just to register, even if I hadn't got a care worker in the end.

Mr Schmid did not wish to pay merely to register and hence did not consider the second offer any further.

During this process of acquiring information, Mr Schmid felt it was important to have references and to hear about other people's experience:

> So I asked a woman who used to work as an outpatient nurse. She knew a former patient who had employed a care worker from that firm in Hungary. I called him and asked how it was going. And he said that he'd had to get used to the food, because he didn't know how to cook and his carer cooked a bit different. And then I asked about her language skills. So he gave the phone to the carer and we talked. I realised that it works very well.

In addition to this reference, Mr Schmid contacted another person that had used the service of that agency. With these two references, he felt confident about trying a live-in care arrangement: 'Then I thought, why not? So when I had the contract, I said, why not? If you're not happy, then the person will be replaced in ten days'. In May 2011, he found a care worker from Hungary with the help of said agency. She stayed for three months until the end of July and worked for another four months from September until January. In August, another care worker replaced her for one month. The care arrangement ended when Mr Schmid's wife passed away in January 2012.

Looking back, it was not a light decision for Mr Schmid to arrange for help, and it took some time before he did. It was not until he felt that caring for his wife had become a burden that he was no longer able to bear that he started to look into live-in care:

> It had gotten more and more troublesome. You had to help with diapers, she had to be showered, given food. She just couldn't do anything by herself. . . . And then I thought, I can't do this on my own anymore. I'd been taking care of her since 2006. In the beginning it wasn't that bad, but gradually it got worse and worse. . . . It had gotten too much, and I just couldn't do it anymore. You have no freedom, nothing, except for holidays [when she stayed temporarily at an old age home]. And I talked to the doctor, and he said, if I take my wife to an old age home for good, and I go visit her every day, and I see that this isn't right, that isn't right. . . . I mean, I'd already seen that when I took my wife to a home for my holidays. When I got back and picked her up, she was wearing dirty jeans and a dirty t-shirt. But there were clean trousers and a pile of clean t-shirts right there! That's annoying. Every day you'd go visit her, and you'd be annoyed. And if you only went once a week, then you'd feel guilty. So I told myself, it has to be okay for both of us, for her and for me. This way, my wife could stay home and I had some freedom. And if it hadn't worked with one carer, I would've got a second carer from this organisation.

Mr Schmid did not regret his decision. On the contrary, he was glad that his wife could stay at home until she passed away. 'She never had to suffer, and she never had to go to the hospital or to an old age home, except when I went on holidays', he emphasised. Moreover, he explained that he would not have hired a live-in care worker without the help of an agency:

> I know someone, they have two Polish carers, not through an organisation, but employed directly. They're two sisters who switch shifts. I'd never do that. Firstly, because of the permits: You have to have a work permit. Then there's social security, old age insurance, unemployment insurance, then the benefit plan. This is all really complicated. And if it doesn't work, then you can't just send her home. And you also need a second care worker who's also good.

For Mr Schmid, it was decisive that he did not have to organise the work permits of the care workers or register them with social insurance himself. In addition, the possibility offered by the agency to change care workers if the relation between Mr Schmid, his wife, and the care worker did not work well was reassuring and decisive to trying a live-in care arrangement through an agency.

Mr Schmid had gathered a substantial amount of information about live-in care arrangements in general and knew of many other peoples' experiences with it. For example, his son and daughter-in-law also employed a care worker from the same agency to care for his daughter-in-law's grandmother. Moreover, he was well informed about price differences between agencies:

> There's an agency in Switzerland, but you just can't afford them. My neighbour had one. They flew one in from Hamburg, she took a taxi to the house, and after two weeks, another one came. That adds up after a couple of months! And then he changed to this agency [the one Mr Schmid worked with] too.

Overall, he reported a very positive experience and was eager to share them with other people interested in a live-in care arrangement with the agency he used:

> Everyone in the neighbourhood . . . I've placed two here, well, 'placed'. I'm asked about my experience all the time. And a woman from Geneva called me. She was looking for help for her mother in Berne. So I explained it to her. Later she called me and told me that everything was great. Then here, a dentist in a village nearby hired someone for his father, and it seems to be going great. And then someone in Zürcher Oberland, they were looking for a carer for their parents. They called me and asked if they could use me as a reference. And down there in the village, by the bakery, they also had someone from the same agency.

Mr Schmid continued to share the experience he had gained of the agency and to give references to other care recipients and their family members who, like him, were looking into a live-in care arrangement. Hence, Mr Schmid became a sort of mediator between individuals interested in live-in care and the agency, which shows how information on live-in care circulates in the community.

The reasons and situations that lead care recipients to organise live-in care are diverse and complex. Nonetheless, Mr Schmid's story provides some insight into the process and the difficult circumstances through which he came to choose a live-in care arrangement. His experiences resonate with elements that Van Holten, Jähnke, and Bischofberger (2013) find in their study on care recipients' family members' backgrounds and decisions to arrange live-in care. Those authors found that the family members had a strong desire for individual care arrangements in their own households, for stable care relief, and a need for security. In addition, the affordability of live-in care was critical to their decisions to organise it (Van Holten, Jähnke, & Bischofberger 2013). In this sense, Mr Schmid had developed similar needs and desires for care relief. The main options he had considered before he looked at live-in care were either to continue to care for his wife himself, which he had been doing for several years, or to take his wife to an old age home. He perceived both options as emotionally strenuous and did not feel comfortable with either. The home care package with the agency thus presented itself as a viable solution to his dilemma. Moreover, his story shows the crucial role that home care agencies and their packaged care services played in his decision. After obtaining references for the agency, the possibility of replacing a care worker, not having to take care of the administrative workload that comes with the formal employment of a worker, and the affordability of the agency's offer were essential to his decision to hire a live-in carer.

The development of home care agencies

Although care agencies have been placing live-in care workers from Eastern Europe in countries such as Germany and Austria for quite some time (see Bachinger 2009, 2010; Krawietz 2010, 2014; Österle, Hasl, & Bauer 2013; Bahna & Sekulová 2019), most care agencies in Switzerland have only arisen since the extension of the free movement of workers to the new accession states in 2011. In this sense, the development of home care agencies in Switzerland arguably represents the expansion of an existing idea to a new geographic area. Indeed, three of the care agents I talked to – Daniel, Nicolas, and Philippe – all started their businesses in Germany and Austria before they expanded their services to Switzerland. However, the majority of the care agencies interviewed had less knowledge and no infrastructure beforehand, as they started their businesses from scratch in Switzerland. How do care agents come to start a business in live-in care in Switzerland?

Insights into care agents' backgrounds

'I always say that we are children of a lost generation', Dora, owner of a one-woman agency, said while laughing, 'you know the communist system. That's how we lived. (. . .) In those ten years, when I went to school and did my education, that was all in a communist system. Only later, the European Union and democracy and everything came'. Having grown up in the communist education system in Slovakia, where education was centrally controlled but generally of high quality, Dora followed a relatively straightforward passage from school to work. By the young age of 18, Dora had already finished her training as a nurse. She did not want to obtain further education. She wanted to work. After half a year in a hospital in Prague, she gained her first experience in Austria as a live-in care worker for four years. Later, after her diploma was recognised, she started to work as a nurse in Switzerland. Dora loved her profession as a nurse. While working in a rehabilitation clinic, she noticed that demand for care work was increasing and that, although there were already institutions for home care, such as outpatient care firms (*Spitex*), she thought that their services were often very expensive and did not cover the demand for home care sufficiently. In 2013, she decided to start her own employment agency for live-in care work.

Care agent Anthony started his business because of his wife, a former live-in care worker, whom he met in the household of a patient that he treated as a physiotherapist. The agency she worked with in Slovakia was looking for a new contact person in Switzerland. Hence, for Anthony, who was already operating a physiotherapy practice, it was social contact that led him to start a new business in addition to his existing one. The reason that care agent Mattea started offering live-in care was that she had moved to Slovakia to be with her partner and felt bored. She started looking for a new hobby or job. Two factors were essential for her agency's launch. The first was that Mattea could draw on the social and business network of her partner, who was well connected in the region. Secondly, she could collaborate with another business partner she knew from Switzerland, who had studied business administration and knew how to create a business plan. It was her business partner's first business project after graduating from business school.

Michael, owner of a placement agency, came upon the idea of launching a care agency through his personal experience with his mother-in-law and social contacts. When his mother-in-law needed care and did not want to go to an old age home, an acquaintance told him about a Slovakian woman who could work as a live-in carer. She stayed for eight years. Subsequently, friends and acquaintances approached him, asking how they could contact other live-in care workers. With the help of a friend's wife. who was connected to care workers in Slovakia, he started his own care agency. Moreover, Michael already had extensive experience of labour placement, as he had worked as a labour broker recruiting migrant workers to the Middle East for many years before he moved back to Switzerland.

The biggest contrast to Dora's one-woman agency in my empirical data is the example of the agency that Andrea and Melvin work for. It is part of a global franchise system that was invented in the US. 'There are investors here in Switzerland; they wanted this business here, so they paid the owners royalties to secure the business's exclusivity here. In return, they have the right to further develop the agency under certain conditions', Melvin, the managing director, explained to me. In this sense, the agency has an exceptional position, as it could build on knowledge from the US thanks to its franchise system. The agency had 21 branches all over German-speaking Switzerland at the time of my fieldwork and specialised in live-in home care. The placement agency that another of my interview partners, Kurt, worked for as a managing director is one of these branches. Some of the branches are independent agencies, whereas some are owned directly by the main business. 'The goal is that as many branches as possible belong to us' said Melvin. Although retired at the time we spoke, he was a former medical practitioner and often a guest lecturer at a university on the topic of demography. 'If you work with demography and ageing, then you'll also come to know about this business', was his answer when I asked him how he came to work with the agency.

These examples are intended to give a brief insight into the diverse backgrounds and reasons that led some of the interviewed care agents to start a new business in 24-hour care. They include social relations, professional background and knowledge in the health sector, general labour brokering experience, and personal experience with elderly care. The starting points of the care agents also differ widely, with some completely new to the domain and others building on existing business structures and knowledge of live-in care.

A volatile and dynamic landscape of care agencies

The scope of the live-in care market is difficult to grasp, in part because the Swiss government lacks means to systematically collect data on live-in care services. The authority responsible for live-in care services is the Department for Employment and Placement, part of the Swiss State Secretariat for Economic Affairs (SECO). The difficulty of estimating how many agencies actually operate arises because the Department is only able to collect data on registered employment and placement agencies in relation to housekeeping services. In 2015, that list contained around 60 companies (Schwiter, Pelzelmayer, & Thurnherr 2018). However, housekeeping services are broadly defined and can include cleaning services and pet-sitting services. Therefore, it is not clear how many of these services actually specialised in live-in care.

Moreover, similarly to Truong, Schwiter, and Berndt's (2012) findings in their study of the home care market in the City of Zurich, I find a diverse and dynamic landscape of care agencies. There are one-person businesses, many smaller firms with two to four employees, and a few larger-scaled agencies with more collaborators. Over the course of three years, from autumn 2013 to 2016, I observed how new companies appeared and disappeared on the Internet and through other forms of advertisements. Sometimes they resurfaced with different names, or the ownership and management changed. For example, in November 2013, I visited a care agency called Seniorenzuhause (not part of this research sample) as part of a guided information tour on migrant care workers organised by the Zurich Asylum Organisation (AOZ). In 2014, the agency changed ownership.

The care agent rejected my request for an interview because he was in the process of selling the agency and switching to the placement of more highly skilled workers in the health sector. Moreover, many of the care agencies whose staff I interviewed underwent changes. Two of them changed management during my fieldwork. Another two changed their names: one of them after bad publicity in the media, the other mainly for marketing reasons. Another one changed ownership. One of the business owners, Anthony, temporarily stopped new placement activities while awaiting regulatory developments, and Livio, the owner of an agency that recruited from Poland, shut down his business completely shortly after our interview. Hence, the market is characterised by an lack of transparency on the provider side with a volatile business landscape.

The dynamic characteristic of live-in care services is similar to the development of domestic employment agencies in other places. In the US, for example, Hondagneu-Sotelo (1997) found a plethora of agencies in the business phone directory and in newspaper advertisements in Los Angeles in the 1990s. In her observation, only a few high-volume agencies had more established structures, whereas many smaller agencies that she characterised as 'fly-by-night' agencies consisted of people with telephone access that had started their businesses 'in a corner of their bedroom' (Hondagneu-Sotelo 1997, 4). In her study on live-in care services in Germany, Krawietz similarly documented a sample with many smaller agencies consisting of one to two people and medium-sized care agencies with four to seven collaborators. She also found difficulties in identifying the number of agencies, stating that the Internet sometimes still contains information on businesses that no longer exist, and that some businesses operate under several names and websites to increase their web presence (Krawietz 2014). In addition, the number of agencies is difficult to assess, as they often operate transnationally and are located in the recruitment countries. Hence, in addition to the fact that there are no official statistics on live-in care services, a comprehensive assessment of the scope of the live-in care market is complicated by its dynamic landscape. Since starting a live-in care service is not a capital-intensive business and does not seem to require a lot of know-how in comparison to other industries, it seems that many of the businesses are small players trying their luck in a new market.

Access to home care agencies and the Internet's role

The locations of the Swiss-based care agencies and the premises from which they operate are as diverse as the care agencies themselves. The agencies included in my research were scattered around Zurich, north-eastern Switzerland, and central Switzerland, with one agency in the Canton of Berne. Interestingly, the majority of them were in rural or suburban areas. In total, I visited four agencies at their official offices in German-speaking Switzerland and two agencies whose offices were integrated into other offices, as the care agents' main activities were in other businesses, such as physiotherapy and student language exchanges. Outside Switzerland, I visited one agency in Slovakia. The rest of the interviews were held outside official office premises: One interview was conducted in the private apartment of the care agent (as a one-person business, Dora operates everything from her own home), another at the University of Zurich, three in a restaurant, and two more on the phone and on Skype (of which one agency was based in Switzerland and one in Hungary).

The various places in which I held conversations with the care agents and their employees are interesting, as they tell us something about how the care market works. Firstly, it partially reflects the care market's early stage of development, one in which care agents explore the potential of their new services. The few more established and larger agencies have set up infrastructures and regular offices, but the majority of smaller agencies work from more informal workspaces. Secondly, it indicates the importance of the Internet and the agencies' web presence, which seem to play a much larger role than the accessibility of a physical office (see also Pelzelmayer 2016). I had the impression that none of the offices were set up so that potential clients could visit them, but simply as a working space in which to operate the activities. And indeed, when I suggested to Anthony that the local people would probably know him well because of his physiotherapy practice, he explained to me that the local network would be irrelevant for his home care agency: 'It isn't a local business, it covers the whole of Switzerland, but especially the German-speaking area'.

Whereas before the advent of information and communication technology, places of businesses and access to them were determined by people's 'ability to move from one place to another via transportation' (Aoyama, Murphy, & Hanson 2011, 60), businesses such as home care agencies benefit from what has been called a 'shrinking world', a term often used to describe the development of transport and communication technologies and the consequent changes in the perception of distance (Allen & Hammnet 1999). The question, then, remains whether interested parties have access to the Internet and sufficient know-how to find live-in care workers on the web. Since Switzerland has developed extensive digital subscriber line (DSL) coverage (above 98%) in the last decade (Díaz-Pinés 2009, 1095; Götz 2013, 10), and given that most of the time it is family members, often partners or daughters, rather than care recipients themselves who arrange care (Wigger, Baghdadi, & Brüschweiler 2013, 86–87; Van Holten, Jähnke, & Bischofberger 2013, 26), neither limited access to the Internet nor a lack of familiarity with its use is likely to form a real obstacle to obtaining live-in care services offered online.

These findings correspond to what Krawietz (2014) found in her study, in which the location of an agency does not matter much, as agents operate without face-to-face contact with potential care recipients and where the relationships between care recipients and care agencies are rather anonymous and fragile. However, it is very different from what Wee (2019) found in her elaborate study on domestic worker agencies in Singapore, where brokerage of live-in domestic workers flourished in the 1990s – when the Internet played rather a minor role – and is now well established and more regulated and institutionalised than in Europe. In Singapore, the physical location of the offices still seems to matter, as it is a common norm for employers to visit the agency, establish a trusting relationship with their agents, and set up interviews with prospective domestic workers. Hence, agencies need shop fronts and sales staff to 'cultivate specific relationships with employers and workers' (Wee 2019, 27). However, in Southeast Asia, the situation differs from Europe, as domestic workers are usually already brought in by the agencies and ready to work. They are often present in the shops and available for face-to-face interviews with prospective employers, waiting to be employed. In contrast, care workers in Europe usually travel to a household in the country of employment after being hired. Thus, the introduction of information and communications technology (ICT) seems to make a difference in the way live-in care markets develop and in how trust relationships are understood and established between agents and care recipients and their families. In our case, the use of the Internet facilitates access to live-in care services from anywhere within Switzerland without requiring physical mobility of care recipients or their families.

Live-in care as fast-growing business?

At the time I visited Mattea's agency in Slovakia in July 2014, the team had just moved to a larger office, as it was expanding. The new office was not well equipped yet, and the Internet connection was not working well. When I arrived, everyone was quite stressed. It was a Monday, and the team was busy closing deals with clients who had had time to think about their offers over the weekend. Daily operations such as recruitment, client acquisition, and client care had to carry on while moving office and installing new infrastructure. The team consisted of seven collaborators: two owners, two sales employees, two recruiters, and one marketing employee. Two more employees were about to start in August. At the time of my visit, their social media site had about 600 to 800 followers, and the agency had gathered a pool of approximately 300 potential care workers. Two years later, in 2016, the number of followers had more than tripled to roughly 3,000 followers.

The overall picture back in 2014 seemed to be that the company was doing well. Client acquisition was growing rapidly. Within one year, Mattea's agency had acquired around 100 care recipients that were still in active contracts with the agency. At the time of our interview, she reported that the agency was closing an average of about 16 new deals per month. Moreover, Mattea stated that she could barely keep up with that growth in developing internal processes, such as back-office procedures. 'I never expected that it would go so well!' she said. In similar vein, Melvin, general manager of a large placement agency, stated that his agency was growing at more than 20% per year, so quickly that he did not have time to pay attention to any competition. It is statements such as these that suggest that the home care market has been booming, as often portrayed in the media.

However, other voices, in both my empirical data and research (see Schwiter, Pelzelmayer, & Thurnherr 2018), complicate this simple picture of sudden growth in home care. Care agent Anthony emphasises that home care is not by its nature a fast business and that building reputation is crucial and takes time:

> It's a business that you really have to build. It is not a business that makes you rich just like that. I have to split the income that I earn with my partner in Slovakia. This is a business that you have to run properly and with responsibility over years to build a name.

On average, he received three to five requests per week, and that was after the initial launch phase when the business was already more established and running. In total, he had placed care workers with around ten care recipients. Similarly, around half of the care agencies I talked to were operating on rather small scales, some of them struggling to acquire clients.

While individual agencies, such as the larger ones with franchise systems and the ones in the low-cost segments operating from abroad, made profits from the beginning of their business and continued to grow substantially, others that I talked to had struggled to become established. Here, the term 'larger agencies' refers to those that had acquired and were managing more than 100 clients, such as Mattea's, Daniel's, and Melvin's. Small-scale agencies have up to 20 clients. What is interesting is that the larger and more quickly growing agencies seem to have made substantial investments in marketing and online advertisement. Their agencies are easier to find using online search machines, and their webpages seem more professional than some of the smaller agencies. Anthony reported that the numbers of requests for live-in care were apparently much higher when he invested in online advertisement and that it had dropped sharply after stopping this. However, not all care agents are willing to pay what they consider to be relatively high sums of money for online advertisement and for the design of their webpages. Pascal, for example, complained to me about the high costs of online advertisement, stating that he did not want to spend that kind of money. Instead, he was pursuing alternative methods, such as personal contacts and business collaborations, to market his offers.

Moreover, the picture of growing businesses becomes more nuanced when focusing on their internal processes. Although the growth of Mattea's agency sounds like a success story from the outside, the process had not always been smooth, and the agency struggled with numerous unexpected problems and obstacles in its business development. Following one particular incident that occurred two months after starting, the young entrepreneur was on the brink of shutting down the business. The daughter of a care recipient had called the agency to inform them that the care worker they had placed had supposedly been drinking. According to Mattea, the agency had suggested replacing the care worker, but the client had decided to continue the arrangement. Three days after that phone call, the care worker fell down the stairs with the man she was taking care of. He broke four ribs and passed away because of his injuries. 'At that point, it wasn't clear whether we would continue this business, whether we wanted to take responsibility for cases like these', Mattea remembered. 'But it's better now. We've gained experience, we've been learning continuously for a year. We learned some expensive lessons. We had to improve everything'. Retrospectively, Mattea thinks, she has gained a lot of knowledge since she started the business:

> I'd do everything differently. We've always employed the right people. But everything else was just not fully developed, even now, not ready for the market. When we started marketing, we didn't even have any contracts ready. We'd anticipated about a quarter to half a year for planning, and then we decided to just start with the advertisement. The homepage was up and running, and everything else was just learning by doing. That cost us clients in the beginning. They were with us and then, understandably enough, they left us again. We learned from that.

Another characteristic of the market is that the businesses seem to be inherently subject to fluctuations. As Daniel, owner of a placement agency, said,

> we're growing constantly. But, let's say, we have 60 customers at the beginning of the month. And then at the end of the month, we have 62. In the meanwhile, we lost maybe ten customers, but gained 12 or 13 new customers.

Mattea added that fluctuations such as lower requests during holidays are normal. She had even seen that the agency had gained around 40 care recipients, and then the number had dropped to around 15 before it started to grow again. She also reported a significant loss in care recipients during winter, saying that many passed away. Thus, it seems that care agencies need longer than some other industries, such as the food and beverage industry, to assess whether their business model works.

What is striking is that, according to almost all of the care agents interviewed, the number of potential care workers is many times higher than the demand for care services. Dora, for example, explained:

> There's a lot of interest [from care workers] in Slovakia and other Eastern European countries. That's just there. I don't have to do much. What I have to work for is to find families here. That's a lot more work. People are less trusting here than in Austria.

At the time we held our interview, in October 2014, she had been officially running her agency for a little more than one month but had been active informally before her official launch. Within around five months, she had received around eight requests, and she had placed care workers with four families. Two of her care arrangements were still active at the time we talked. The other two arrangements had ended because in the first case, the care recipient had passed away, and in the second, the family had decided to use an alternative arrangement. Care agent Pascal reported about five requests per month from potential clients, in contrast to about 50 requests from care workers who would like to work for him. These comments indicate an imbalance between the supply of live-in care workers and the demand for live-in care services (see also Truong, Schwiter, & Berndt 2012).

While the care agencies experience very different forms of growth and struggles in developing their businesses, there seems to be general consensus on the potential of the 24-hour care market among all care agencies. The basic assumption seems to be that the live-in care market is growing and will continue to do so. Correspondingly, many of the care agents interviewed reported big plans for the future. Dora, with her one-woman business, is convinced that her business will develop because, in contrast to others, she would be offering 'fair and legal services'. She planned to be able to give up her job as a nurse in a rehabilitation clinic and to work as a care agent full time within three years: 'I'm a hundred per cent sure that there's potential. Because Austria and Germany are already right in the middle, Austria for at least 15 years, Germany seven or eight years. That's the future'. Dora was not the only one who related care market developments in Switzerland to those of its neighbouring countries. Mattea talked at length about expanding recruitment to Eastern European states such as Bulgaria and Romania and placing care workers in other German-speaking countries and the Netherlands. Her vision was to become market player 'number one' in the whole of Europe. Thus, while visions are grand and expectations for the market are high, growth and business development ments seem to be anything but frictionless, and the market is uneven, with small agencies struggling to acquire clients at one end and fast-growing investment-intensive agencies at the other.

Conclusion: live-in care made accessible

Live-in care in Swiss households is by no means a new phenomenon and has existed for a long time. Hence, migration for live-in care work in Switzerland, mainly in the form of informal arrangements, existed despite restrictive migration regulations before the introduction and extension of the free movement of workers agreement. However, changes in migration regulations have led to new private businesses that place care workers in households, thus changing the characteristics of migration for live-in care work in Switzerland. As this chapter shows, these care agencies can function as facilitators of would-be care recipients' access to live-in care arrangements. Although it is possible for private households to hire workers directly, and I heard of and encountered several instances of direct employment during my research, without a private contact that can connect a would-be care worker to a private household as employer, the obvious and easiest way is to contact a care agency.

Overall, it seems that the barriers to attempting a live-in care arrangement have been lowered with the development of a diverse landscape of agencies offering packaged live-in care offers. This development has been facilitated by progress in information and communication technologies, enabling care recipients and their family members to access information about home care services in their own homes and organise care arrangements without having to travel. Particularly for care recipients without prior experience or private contacts, it is easier to access and arrange live-in care online. Not only can care recipients outsource organisational expenses via home care packages, but they can also compare information about different all-inclusive services online. Navigating this process and collecting references might take considerable time before a care recipient or family member actually books a service. As can be seen from Mr Schmid's story, care recipients and their family members exchange information and pass on references to each other. It is not uncommon for care recipients to test the services of different agencies and change from placed to unplaced care services and vice versa. Various care agents have told me how they have apparently taken over other care agencies' care recipients. I also heard several stories from care workers and care agents that employers decided to formalise previously informal care arrangements with the help of an agency or to continue a placed care arrangement without the care agency's knowledge. Hence, in addition to the services available online, information on experience and prices circulates in the community.

The new care market provides not only a new service to an ageing society but also lateral entrance for business starters even without previous knowledge or experience in the field of care work. To start their businesses, care agents draw on both their own personal networks and their professional relations and experience.

Accordingly, care agents have very different starting points in terms of knowledge about business development in the live-in care sector and accumulate experience in learning-by-doing processes. Consequently, the development of home care businesses has not been frictionless. While all care agents have grand visions for their businesses and agree on the potential of the market, they have experienced very different difficulties and growth rates in their start-up journey. To understand why businesses differ so much in their development, we also need to understand how these agencies are regulated. The next chapter provides insight into the regulatory contexts of live-in care agencies.

4 How regulations matter for care agencies and care workers

In most countries where employing domestic workers in private households has become common, employers usually directly employ domestic workers and, hence, figure as employers. The Swiss case differs. Some of the care agencies in Switzerland not only recruit and place migrant care workers but also act as employers. Care agencies offering live-in care services are regulated by the Recruitment and Hiring of Services Act (AVV 1989), which distinguishes between two different types of agencies: *Personalvermittlung*, which is here translated as employment agency, and *Personalverleih*, which is here translated as placement agency. Figure 4.1 illustrates the difference between the two forms of business.

Employment agencies act as brokers and collect a fee for their services, so the care recipients and their families act as employers. Much of the global literature on the recruitment of domestic workers refers to these types of agencies, which are often called recruitment agencies (Hondagneu-Sotelo 1997, 2000; Wee 2019; Maher 2004; Liang 2011). In contrast, placement agencies employ care workers directly and place them in households in Switzerland. Consequently, the agencies function as the employers. Thus, the employment relations that predominate in Switzerland differ from those described by the literature on domestic workers in Asia, America, and the Middle East, in that care recipients and their family members have limited responsibilities; the placement agencies bear the main responsibilities for paying wages and ensuring compliance with local working conditions. Although care recipients and their family members supervise and direct day-to-day work, they are not formally responsible for the legal and administrative aspects of employment.

Therefore, Switzerland presents a more complex case than other countries, with various employment situations. In addition to direct employment of care workers by care recipients mediated by employment agencies, care workers can be in triangular relationships, with placement agencies as their legal employers and care recipients as their clients. In our sample, 8 out of 13 agencies operated as employment agencies and five as placement agencies. The question then arises, how are these agencies regulated? How do these different arrangements affect care workers' working conditions? This chapter first locates care agencies as part of a wider global increase in employment and staffing agencies. It then explores how agencies are regulated and how these agencies deal with regulations. This is an important basis for understanding how different kinds of regulations and

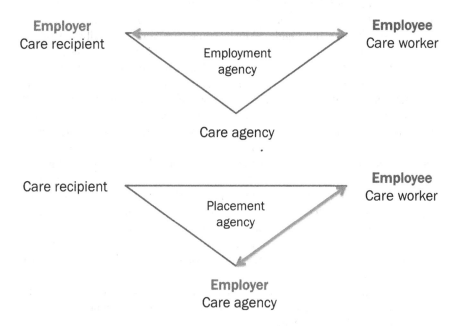

Figure 4.1 Difference between an employment agency and a placement agency.

business practices interplay and, as I will argue, shape live-in care workers' working arrangements and working conditions.

Care agencies as part of the global rise of employment and placement agencies

The global rise of employment and placement agencies, often referred to as staffing agencies in the literature, and its implications for social transformation have been extensively researched, notably in labour geography and sociology. Although labour market intermediaries have existed since the 1950s, their number has increased significantly since the 1980s (Ward 2004). While orthodox economic approaches understand labour market intermediaries as neutral agents that efficiently match labour supply with labour demand by reducing information asymmetries (Benner, Leete, & Pastor 2007), a rich and growing literature has focused on labour market intermediaries as agents and emphasised their role in changing and shaping labour markets (Benner 2003; Coe, Johns, & Ward 2012; Peck & Theodore 2001, 2006; Ward 2004). The growth of the temporary staffing industry has been discussed as a neoliberal practice and as a driving force in the erosion of traditional employment relations. The restructuring of labour markets through temporary staffing agencies varies according to geographical context and often leads to more flexible labour markets and a casualisation of employment relations

(Coe, Johns, & Ward 2009, 2012). Moreover, the landscape of employment and placement agencies and the markets have been shown to be polarised between agencies specialising in highly skilled employees and low-skilled workers (McDowell, Batnitzky, & Dyer 2008; Peck & Theodore 1998, 2001, 2002). Thus, the global increase in labour market intermediaries and temporary staffing agencies has introduced divisions and poorer working conditions within the work force.

In her analysis of the origin and development of international employment agencies in Europe, Pijpers (2010) follows Krissman's (2005) call to pay more attention to the labour demand side in relation to migrant labour. She argues that employment agencies, which began as anchors of circular labour migration between Poland and the Netherlands in the agricultural sector, have gradually expanded into other sectors and play a significant role in the circulation of migrants in the whole of Europe. Interestingly, she observes that international employment agencies 'constantly have to reorient their strategy and reprioritise and reorganise their activities' as they are confronted with dwindling labour pools in certain geographical regions (Pijpers 2010, 1087). In their quest to circumvent regulatory barriers and to increase their efficiency, the employment agencies found a solution in a practice known as 'posting' workers (European Commission 2015; European Parliament 2006; Pacolet & De Wispelaere 2016). Workers are 'posted' if service providers win contracts in another country and send their employees there on a temporary basis to fulfil the contract. By transferring the recruitment, coordination, and administration of workers to local offices in recruitment countries, which then post workers to countries in Europe, employers pay lower wages and decrease bureaucracy (Pijpers 2010, 1087). Adopting the term 'migrant flexiworkers' to refer to those that are placed by an international placement agency, Pijpers's (2010) study demonstrates how labour markets are made increasingly flexible in a transnational dimension.

Furthermore, Pijpers's (2010) study shows how labour brokering gains its own dynamic and logic as placement agencies constantly develop new strategies to maximise their efficiency. Hence, employment and placement agencies have not only multiplied in numbers but have also shown themselves to be dynamic, expanding into sectors in which staffing agencies have not traditionally operated, such as health care and IT, and benefitting from global inequalities between recruitment countries and countries receiving migrant labour (Ward 2004, 259). In this sense, the rise of home care agencies in Switzerland is part of a growth in the employment and placement industry more generally and the expansion of this industry into new sectors and markets in Switzerland. In our case, it is the care sector and a new market of live-in care services. However, new markets often entail new regulations and social challenges. How does the Swiss government deal with the emergence of employment and placement agencies selling live-in care services?

Regulating home care agencies

While Switzerland has introduced free movement for workers from the European Economic Area (EEA) countries, it has regulatory barriers on free movement of services. In contrast to Germany and Austria, Switzerland does not allow cross-border services (*grenzüberschreitende Dienstleistungserbringung*) in the home care sector. In other words, while posting is generally allowed and common for live-in care workers within the European Union, it is not allowed for the live-in care sector in Switzerland, which has signed a restricted agreement on freedom to provide services with the European Union (Medici 2015, 113–14). Therefore, the placement of live-in care workers to Swiss households is only permitted by agencies that are officially based in Switzerland and registered as such.

The Recruitment and Hiring of Services Act (AVV 1989) stipulates that agencies that recruit and place employees have to acquire a permit at the cantonal offices for economy and labour in order to operate. It distinguishes between employment agencies (*Personalvermittlung*) and placement agencies (*Personalverleih*). As described previously, the former act as brokers and collect a fee for their services so that the care recipients and their families act as employers. The latter hire care workers and place them in households and, hence, function as employers. Placement agencies are subject to more regulations than employment agencies and have to deposit a sum of CHF 100,000 with the cantonal office of economy and labour to guarantee salaries (AVV 1989, Art. 14; AVG 1991, Art. 6). In addition to permits from the cantonal offices for economy and labour, agencies that send and recruit staff to and from abroad also have to obtain a permit from the Swiss State Secretariat for Economic Affairs (SECO) (AVG 1991, Art. 21).

Employment or placement agency?

In cases where care recipients and their family members use the services of employment agencies, it is the families that act as employers, and therefore, it is their responsibility to carry out the administrative work arising from the employment. The care recipients and family members have to arrange health insurance and social insurance for the care workers, pay salary, issue pay slips, and apply for a residence permit. However, some care agencies start out as employment agencies offering all-inclusive packages, which include not just placement but also salary-paying services and other forms of administrative work. If these tasks are undertaken by the employment agency instead of the families, then the agencies' activities are considered illicit, as Thomas, an employee in the Department for Employment and Placement at the Swiss State Secretariat for Economic Affairs (SECO), informed me during an interview in winter 2015. From the perspective of the cantonal offices for economy and labour and the SECO, Thomas stated, the employment agency would either have to stop its activities or become a placement agency. The authorities responsible at both cantonal and national levels determine whether and when agencies function as employment agencies or should be recategorised as placement agencies.

The line between an employment agency and a placement agency seems to be a fine one. For example, Dora's business is categorised as an employment agency, because she explicitly does not offer to take over the families' responsibilities as employers; she only offers consulting services regarding salary administration, registering with social insurance and related authorities, and registering the care worker's residential status with the migration authorities. Employment agencies that do not explicitly make this distinction between offering advice on administrative work and performing administrative work can be asked to register their agencies as placement agencies. For example, care agent Livio, whose agency began as an employment agency, was told to acquire a license as a placement agency. Anthony's care agency had offered two different services, a basic package and a plus package. The basic package only included the placement, so that the care recipients' families took the role of employer and thus were responsible for carrying out all administrative work related to the employment. If care recipients took up his plus package, Anthony carried out the administrative work for them. He stressed that he prepared all the forms so that all the family had to do was sign. In the future, if Anthony resumes his service, he does not plan to distinguish between basic and plus bundles anymore. Instead, administrative work would be included from the outset, and the family would pay a lump sum. During our interview, it emerged that Anthony was not aware that whether he has to obtain a placement agency permit depends on whether he markets his offers as administrative work in consultation form or as included service. Having already obtained a license as an employment agency, the question is whether the authorities would come to know should his activities require a placement permit.

Dealing with new and unlicensed home care businesses

In practice, according to Thomas at the SECO, the Recruitment and Hiring of Services Act is not very well known. 'Even in other sectors, businesses often don't know about it and don't realise that they actually need a permit', he explained. Although there are penal provisions for businesses that deliberately operate without a permit, at the time of the interview (December 2015), the SECO had not yet prosecuted a live-in home care business. Instead, Thomas stated, it would inform such a care agency about the regulations and either oblige them to obtain a permit or ask them to stop their operations.

To receive a permit, agencies officially have to be registered in the commercial register and hence, be based in Switzerland, and they have to have an appropriate office (*zweckmässiges Geschäftslokal*; AVV 1989). That means that the registered office cannot consist of a private apartment, unless there is a separate office within the apartment and the building has correct signage, Thomas clarified in our interview. In that case, the SECO would accept a private apartment as a registered office, but more readily for an employment agency than for a placement agency. Moreover, the manager has to possess a residence permit, and department managers (*Fachbereichsleiter*) have to possess appropriate occupational (*fachgerecht*) skills and ensure professionalism in business operations (AVV 1989, Art. 13).

Finally, businesses have to provide valid employment and placement contracts, which are examined by both the cantonal offices for economy and labour and by the SECO. Contracts vary individually in working hours, but generally care agencies are expected to stipulate the numbers of normal working hours, on-call time

(time during which care workers have to be available to work in case they are needed, and hence cannot leave the house), and time during which care workers do not have to be available and can leave the house. Thomas explained that the SECO has accepted the practice of working 14 days in a row without a rest day. The purpose of this practice is to enable care workers to rotate on and off their jobs in two-weekly cycles. However, the number of hours divided between working, on-call time, and free time varies from case to case. According to Thomas, it would not be possible to set a standard. Asked about the remuneration of on-call time, Thomas explained that the SECO accepts a rate of three to five Swiss francs per hour. This norm, according to Thomas, seems to have been based on an existing practice in the Canton of Zurich. A decision by the Federal Court stipulates that the standby service of workers has to be compensated, but it does not stipulate the amount of compensation (*BGE 124 III 249 E.3* n.d.).

If the SECO learns of a home care agency that does not have the required permits yet, Thomas stated, the SECO first checks the agency's background. It examines the agency's Internet presence, checks whether it is registered in the commercial register, and seeks other sources of information such as brochures. Subsequently, the SECO contacts the agency directly and informs them about the permits they require. 'Often, they don't know about it [the requirement of a permit], and it's difficult to prove that they knowingly operated without it', Thomas said.

Dealing with illicit home care agencies

In Thomas's experience, businesses in Switzerland usually yield after such contact and start the process of acquiring the permits they need, but this is more difficult with businesses based abroad. Care agencies from abroad are officially not allowed to place in Switzerland (AVV, Art. 30). Most of the time, as Thomas explained, it is other businesses that report illicit placements by agencies abroad to the SECO, or private individuals that ask whether they are allowed to enter a contract with these businesses. Between the introduction of the free movement of workers for EU accession countries in 2011 and 2015, when I interviewed him, Thomas reported that the SECO contacted approximately 50 to 60 businesses based abroad which, he suspected, were active in live-in care. The SECO usually contacts these agencies and informs them that to be able to place in Switzerland, they have to register in and operate from Switzerland.

According to Thomas, most of the businesses the SECO contacted abroad answered that they did not know about the permit requirements or that they had to be registered in Switzerland, since the posting of workers to Germany and Austria is allowed; thereafter, they stop their activities in Switzerland. However, some agencies still continue to place care workers in Swiss households, to the detriment of agencies that are formally registered in Switzerland, as care agent Anthony complained: '[there are] businesses who, despite knowing [they are not supposed to] still operate in Switzerland'. In the course of my fieldwork, I came across at least four care agencies that were not based in Switzerland and yet placed care workers in Swiss households. Two of the agencies I talked to in person, and the other two agencies I learned of in stories from my interview partners. Thomas explained the difficulties in preventing these agencies from placing care workers to Switzerland:

> We tried different measures with them, we reported an offence, but the problem is that the criminal authorities already have enough to do. And it seems they don't like to convict people who are not here, that is, in absentia.

In cases where agencies continue to place care workers in Swiss households even after being warned by the SECO, Thomas explained, either the cantonal offices for economy and labour can issue charges for violating the Recruitment and Hiring of Services Act or the SECO can issue a charge of unfair competition. However, the prosecutor would have to sentence the defendants based on evidence such as existing contracts or information on the agencies' websites indicating clearly that they place care workers without permits. Since private households are unlikely to volunteer to testify, as they too could be liable to prosecution, it would be difficult to collect evidence.

Moreover, in Thomas's experience, convictions that are only punished with fines are usually only pursued if there are agreements between the countries involved, as with traffic fines. With regulations for the recruitment and hiring of persons, he pointed out, there are no such agreements between Switzerland and other countries in Europe. As for the agencies that the SECO was trying to prosecute, the charges against one agency were dropped. With another, the outcome was not yet clear at the time of our interview with Thomas (December 2015). The difficulties, as he explained, are that

> with the rest of the agencies, we assume that they do not place in Switzerland anymore. . . . However, we can't actually verify that. Because they have a web presence that is in German – they can always claim that it is only meant for Germany and Austria. Before, they used to have special links for Switzerland. If they delete those, we cannot trace it.

Navigating through regulations and circumventing regulations

The care agencies' own comments resonate with Thomas's explanations of how care agencies and the SECO interact. Many of the care agents reported that they were contacted by the cantonal offices and the SECO and have acquired information and knowledge in interaction with the authorities. Dora's case is an example of stepwise development towards a business that is considered legal by the state. To start her care agency, Dora explained, she had to take a number of steps. First, she undertook additional training with Cura Viva, the national umbrella organisation of homes and institutions, to have her diploma recognised by the Red Cross as a self-employed registered nurse (*selbstständig diplomierte Pflegefachfrau*). Furthermore, she had to take German courses up to level B2 (upper intermediate level in the Common European Framework of Reference for Languages) and send the course certificates to the Red Cross for her nursing diploma to be recognised. She received the diploma after two years in Switzerland. Subsequently, she applied for a permit for a private employment service (*private Arbeitsvermittlung*) and for a permit for the international recruitment of workers (*private Auslandsvermittlung*), both of which she received after half a year. Finally, she explained that she had to attend more additional training as a personnel administrator. After starting the course, she had to send in the confirmation of her registration, and then she could start her self-employment. She concluded:

> You need a lot of patience. But I have to say; the offices were very open and helpful. If I had a question or filled out a form wrongly, they contacted me and communicated clearly. It took months to get every permit, but eventually it was worth it.

Moreover, Dora received support and experience from the first family that she worked with on the content of the employment contract: 'The first family helped me a lot. I benefitted a lot, in terms of the contracts that I offer for my service. They helped me to design the contract'.

However, not all care agents expressed such a positive attitude towards the cantonal offices of labour and economies and the SECO as Dora did. Nicolas, managing director of a placement agency, compared the situation to Germany, where the posting of care workers is allowed, and complained that agencies have to abide by many more regulations in Switzerland: 'It is so much more diffuse [than in Germany]. You have to get this placement permit; you have to go to different administrative offices. It is so complex; we are still busy dealing with that'. He continued:

> [There are] different commissions that keep us busy. [Some] contracts are accepted in one canton but not in another one. We are still in that phase when we try to satisfy all the different interests and fulfil all the requirements. (. . .) We talk to the cantonal office for commerce and employment. Then they often refer to the Swiss State Secretariat for Economic Affairs, and then they send us back to the cantonal office. Our impression is that nobody really knows or dares to know what is really legal and 100% embedded in the legislation and what is possible. Where is the wiggle rooms, where are the grey zones? So one person says, it is possible to do it this way. But then, whether there will be a lawsuit, a court case, whether what we do will hold in court, no one knows for sure. Even with the employment contracts: There are clear guidelines, but if it's about specific aspects in the design of the contract, they say, okay we accept your draft version and give you permission to proceed, but with reservation, so if an employee brings a charge against you, we can't say how a judge will decide.

As a consequence, Nicolas explained, his agency has decided to restrict their growth and to not take care of more than ten families. 'None of us wants to risk our own livelihood, because we all have other projects too', he said. He wished that there could be more clarity, especially about working time guidelines and resting periods: 'The state has to create clear guidelines so that agencies know clearly, if I behave like this, I'm in a grey zone, or in a legal zone'. Thus, Nicolas criticised a lack of clarity in the guidelines. When I asked care agent Mattea what she would tell new agencies with less experience, she recommended working with a lawyer to gain more clarity. However, she also emphasised that it only helps to gain information on guidelines, but not on how to deal with grey areas.

When Michael decided to start his agency in 2006, he worked for two years with lawyers to find a way to offer live-in care work legally. In addition to uncertainties about working hours and salary negotiations, the free movement of workers had not been introduced for Slovakians in Switzerland at that time. While he was still working out how to design his live-in care offers, his contact in Slovakia proposed placing skilled workers from Slovakia, so Michael started to place certified nurses and doctors in hospitals in Switzerland. Before 2011, the regulations allowed a quota of skilled workers from the EU-8 accession countries. Only years later, according to Michael, did he receive the green light from the SECO for his placement activities in live-in care. However, he complained that his agency would not succeed with their services in live-in home care, as he would not be able to compete with the illegal practices of other care agencies.

The opinions of the care agencies about how regulations might be improved differ widely. Whereas some of the interviewed care agents would like a model similar to that in Germany that would allow the posting of care workers from agencies operating from outside Switzerland, others clearly condemn agencies that legitimise their operations with the argument that Switzerland should sign an agreement for full free movement of services. They argue that the authorities would not do enough to prevent illicit placements from agencies based abroad that undermine the market with low-cost live-in care services. They demand stricter regulations and law enforcement to shut down illicit agencies and thus secure fair competition between the agencies.

Some care agents feel that they are negatively affected by illicit activities and services from agencies operating abroad and by the state's limited options for preventing their activities. Such agencies have evidently gained from these regulations and monitoring gaps. Mattea's agency, which is not officially allowed to operate in Switzerland, has managed to grow substantially by bypassing the Swiss regulations. Having set up her office in Slovakia, she has a great advantage over her competition in Switzerland, as her expenses for office rent, office infrastructure, and salary for administrative employees are much lower. When establishing the business, Mattea had worried about public opinion and the letters she received from the Swiss State Secretariat for Economic Affairs stating that her activities were illegal and unacceptable. Although she perceived the market and the regulations as 'chaotic', she kept pursuing plans to expand the agency's services. Consequently, when I asked Mattea during fieldwork in 2014 about possible tighter regulations on working hours and remuneration for care workers, her response was quite relaxed. For her, more restrictive regulations in Switzerland would be a winning situation, as they would affect her competition in Switzerland more than her agency in Slovakia:

> Switzerland is expensive. For us, it is even better if there are more changes of people [care workers], because we can handle the changes for lower costs than some temporary staffing agency in Switzerland . . . They have many more expenses. It makes a difference in the organisation of permits. For a Swiss employer, it costs 840 francs to organise a B permit [residence permit], if he organises it in Switzerland. An L permit [short-term permit] costs around 40 francs. For us, an L permit costs around 4 Francs. A B permit requires more work; that costs around 50 francs in the Slovakian market. It plays into our hands that we outsource everything to Slovakia. . . . It would be a win–win situation for us [if admission of care workers were regulated more strictly], because the demand will not diminish only because of changing regulations. People are not getting any younger. Every bureaucratic burden plays into our hands. We're not scared at all.

By 2018, Mattea's agency had officially been sold, so Mattea is legally no longer responsible or liable for the agency's live-in care services. She has expanded her activities to general recruitment and brokering of workers for many other sectors and has taken up new activities in real estate business and consulting. However, the question remains whether and to what extent Mattea is still in control of the home care agency's activities. At the time of my visit, her idea was to sell the agency to another firm, founded under the name of one of her employees in Slovakia and in collaboration with a placement agency in Switzerland.

Grey areas

Overall, we can identify two main grey areas. The first is that, even after obtaining all the permits required by the cantonal offices for economy and labour and the SECO, it seems that some uncertainty remains for care agencies about working hours. The contents of the working contracts are defined individually depending on the care recipient's situation and negotiated between the care agency and the authorities. However, due to the nature of care work, in reality many care workers work longer than the six to eight hours a day stipulated in the contracts (see Chau, Pelzelmayer, & Schwiter 2018; Schilliger 2014). The people they look after often require around-the-clock care. Hence, if a care worker decided to file a suit against the care agency to claim reimbursement for their effective working hours, the outcome would be difficult to predict. For example, Pascal was told by the cantonal office of economy and labour: 'You have to see, you are in a grey area if you do this. You have to think about this carefully'. He continued: 'And then we discussed it for about two, three hours. And in the end, they said, well, where there's no plaintiff, there's no judge (*wo kein Kläger ist, ist kein Richter*)'. Hence, permits are not a guarantee that care agencies' contracts will hold in court.

So far, there has been one case in Switzerland in which a care worker sued a placement agency in the civil court of Basel for unpaid working hours and on-call time. She was supported by the Respekt group, and we had the opportunity to participate in the court ruling. The care worker was able to prove her additional working hours, and the court ruled in her favour. The agency had to reimburse additional working hours in full and pay half of the usual hourly wages for every hour spent on call. For three months of live-in care work, the agency had to pay an additional CHF 17,000 (Schilliger 2015). In 2017, the Respekt group hired a lawyer to take on all the cases that are filing complaints against agencies. Most of the disputes are likely to be resolved by settlement before going to court, the lawyer told me in an informal conversation in May 2017. It can be assumed that there are many more disputes than we know of between care workers and care agencies, and that these are resolved in settlement.

In light of this high compensation in the court case, it is no surprise that some of the care agencies expressed uneasy feelings about their activities and are wary of the possibility that a care worker could file a suit against them. For example, for Anthony it was not yet worth expanding his care business as he already had a main business, his physiotherapy practice, which was going well. He had temporarily stopped all his brokering activities by the time we had our interview in 2014. Even though he had obtained all the permits, he felt that it was too risky and decided to wait for further regulations. In this sense, he felt that

> the state has put a brake on it. If I do something halfway incorrectly, I'm responsible, and if in five or ten years there are suddenly huge demands from the authorities that I can't pay, then I'd go bankrupt. It's known that the state (*der Bund*) will maybe work out some official framework. I'll monitor those discussions a little bit, and then, when the conditions change to my advantage, I'll hop on again.

The second grey area occurs if a private household hires a care worker with the help of an employment agency but then does not fulfil their duty as an employer to register with the authorities. The result of this is that their care workers work in informal employment. This chiefly concerns care agencies that market all-inclusive care services for very low prices, such as agencies operating from abroad. During my fieldwork, I encountered some cases in which private households working with employment agencies from abroad registered the care workers and others in which they did not. However, it seems that this grey area also concerns agencies based in Switzerland, much to the annoyance of some of the interviewed care agents. They emphasised that they try to run their businesses legally: 'Others earn money, their main earnings, with that. They slip into semi-illegality. They'd never admit it. I know exactly what kind of excuses they have, saying that everything's okay in the legal domain. Yeah, right', Anthony complained. For him, it was important that the families meet their responsibilities and that he tried to monitor whether they do so or not, as he wants the care workers to work with him again in the future. 'With some families it works well, with others not at all. . . .

Many agencies that work the way I do [as an employment agency] shirk their responsibility'. What Anthony described here is a grey area in which employment agencies place care workers while not bearing responsibility for the consequences of the care workers' working conditions. Although private households are by law obliged to undertake the corresponding administrative work for any formalised employment, not all private households as employers do so and there seem to be no monitoring mechanisms to ensure compliance.

Home care agencies' roles in care arrangements and working conditions

This diverse landscape of agencies has produced a wide range of placement practices that have resulted in various working arrangements, from informal to formal labour and from undocumented residence to long-term residence permits. In general, it seems that placement agencies usually register the work and residency of their care workers with the authorities. The reason for this could be that they are subject to more regulations than employment agencies and as employers are liable for correct conditions of employment. Typically, the working arrangement organised by placement agencies consists of two care workers that rotate every few weeks in the same household and that go back and forth between their workplace in Switzerland and their home in another country. One of the interviewed placement agencies, for instance, has official permission to employ their care workers for a continuous length of two weeks without a free day and hence rotates care workers every two weeks, as care agent Andrea explained and as Thomas confirmed was accepted by the SECO.

Saving costs at the expense of care workers

There are also care workers who work informally and whose residence is not registered, especially when directly employed by the care recipients and their families. This work arrangement can be difficult for live-in carers, as this quote from Anna shows:

> I did not like that. I did not want to work illegally. I asked the family and they took a very long time. Finally, after around two years, they organised my registration and I have an official residence permit now.

When asked why the family took so long, she responded: 'Because of money'. Anna initially came to work in Switzerland with the help of a private contact. In this case, it is the responsibility of the care recipients and their families as employers to organise a residence permit for the live-in carers and to register the care work. Generally, employers have to pay more for formalised employment than informal employment because of taxes, health insurance, and social insurance, such as old-age provision. Above a certain threshold, employers and workers also have to contribute to an occupational benefits insurance (*Berufliche Vorsorge*), which increases the costs further. Hence, in contrast to informal employment, formal employment relations can cost up to 'several thousand Swiss francs' more, as care agent Anthony observed. The higher costs may well be why private households choose informal arrangements or delay the administrative process, even if the administrative formalities are sometimes prepared by the placement agencies and ready to sign, as care agents Pascal and Anthony pointed out. Both these care agents stressed that they try to ensure that the employers follow the procedures required for formalised employment. Not all care agents I talked to stressed this issue. Thus, it may be that the care agents' attitudes to employers' compliance with the legal conditions sometimes plays a role in care workers' conditions of employment.

In addition to omitting or delaying administrative work to decrease costs, other workarounds involve carefully planning work schedules to avoid paying occupational benefits insurance. Anthony declared, when talking about the difficulties of competing with low-offer employment agencies, that

> almost all agencies and employers reach the limit and ought to pay [occupational benefits insurance]. But they use tricks. They arrange for the women to start their employment in the middle of the month. Oh look! They only earned this much per month [not enough to reach the threshold]. You know, they trick and shift things and I understand all that, but I don't want to do that.

Since controls are difficult to implement in private households, as SECO employee Thomas stated, it seems that it is also easy for employers to falsify pay slips to avoid occupational benefits insurance. 'They did my working contract pretty badly. I work four weeks out there, and I'm home for another four weeks. But the contract says that I work 21 hours a week, continuously', said Sybille. Sybille first worked in Germany before she found a job in Switzerland with the help of a private contact. In her first direct employment in Switzerland, she worked informally for seven years. Her second employer finally arranged formal employment. Although Sybille was generally glad about that, she was also confused and worried about her new situation:

> This gentleman [the care recipient's accountant] told me things, we made the working contract. I agreed. He suggested dividing the wage so that I get it every month. I thought, why not, what disadvantage could it bring me, let's do it. And later I looked it up how this works, what I'm required to do, I looked it up a bit on the Internet. And I was a bit surprised and realised that we have serious omissions.

Sybille's employer had her sign a false contract, which stated that she would work continuously and part time, instead of stating that she was working full time every second month. This way, instead of paying Sybille the full salary of one month's work every second month, he pays her half of her salary of one month's work every month. As a consequence, the employer does not have to pay occupational benefits insurance for Sybille. In addition to not receiving occupational benefits insurance, which Sybille was entitled to, her employer did not inform her either that her registration in Switzerland meant she would have to pay health insurance, which her employer did not pay for her. Since health insurance in Switzerland costs between 200 and 400 Swiss francs per month (Priminfo 2017), it is a significant burden on her income, which she was not aware of and so could not calculate when she agreed to her new employment.

Arguably, care workers that are placed by an agency operating from abroad face higher risks of unknowingly working informally and residing undocumented than care workers working with agencies based in Switzerland. During my fieldwork, I encountered stories about informal employment of care workers by both of the care agencies I interviewed outside Switzerland, in Hungary and Slovakia. Sometimes, the interviewed care workers did not know that the private households had not complied with the legal requirements in Switzerland. By passing the responsibility for registration and other administration of formal employment on to the private households, the placement agencies can save administrative work and costs.

How migration regulations foster short-term placements

The main operating model of some placement agencies, such as Pascal's and Mattea's, is to replace one care worker with another in one household after three months or even less, so that an elderly person is cared for by at least four different care workers during one year. The reason that agencies work in these ways is connected to migration regulations. Within the free movement of persons agreement, nationals from the European Union who work for less than three months in Switzerland only have to register their stay through an online form with the local authorities. Workers who stay longer have to apply for a residence permit and have to pay health insurance in Switzerland. With the model of constant placements of new care workers, placement agencies can offer live-in care arrangements for lower costs. For care workers working with these kinds of placement agencies, it means that they can only work for up to three months in private households in Switzerland and have to wait for the next year to be placed again.

The specifics of a care arrangement are negotiated between the various actors: care recipients, family members, care agencies, and care workers. If the participants agree to a solution, it can result in a new placement. Berta came to work in Switzerland through a care agency based in Hungary. She wanted to rotate at regular intervals, so that she could be with her family in Hungary and because she felt that working for longer periods was harmful to her health. However, the care recipient's family wanted her to stay for six months, take a break for one month, and then have her work again for another six months. Berta finally declined this proposition and asked the agency to place her according to her wishes. Subsequently, the agency placed her in a new household and proposed two-month shifts with one-month reliefs, and that the person on the relief shift would change every time. The reason that the shift partner would change every time, Berta explains, would be because 'nobody wants to sit two months at home and work for only one month'. For the persons on the relief shift, and generally for care workers that are frequently placed in new households, this can be tiring. Having to become acquainted with 'a new environment, a new situation, a new disease, a new person' every time is very exhausting, as Marina, another live-in care worker from Hungary, stressed. Berta explained why the agency prefers a rotation of two-month shifts in a row with a one-month relief: 'It's good for them, they don't have to transport people for free that much'. Later Berta switched to a private employment, but she stressed that if she lost this employment, she could go back to the agency and was sure that they would have work for her.

Finally, one of the interviewed placement agencies employs an entirely different model from most care agencies. The agency does not make use of circular migration but arranges for care workers to stay in Switzerland for at least half a year. It has negotiated permission that a care worker can work for three and a half days in one household so that at least two care workers have to change shifts in one week, as Yvonne, the managing director, explained to me. However, depending on the health situation of the care recipient, the agency also uses arrangements in which care workers only stay one or two days. During their off-times, the care workers live in an apartment that is provided by the agency. However, out of our entire sample as well as all the agencies we heard of, this was the only one that offered live-in care services without employing short-term circular migration. All the other agencies rotate their care workers on weekly and monthly bases.

Conclusion: flexible working arrangements in live-in care

The emergence of care agencies has kept cantonal offices of economy and labour and the SECO busy controlling their brokering activities, preventing unfair competition between the businesses, and ensuring compliance with local working conditions. The live-in care market in Switzerland is young, and several grey areas are subject to negotiation and still remain unresolved. I demonstrated that even after obtaining all the permits required to operate a home care agency in Switzerland, certain insecurities still remain for the agencies. Moreover, some employment agencies place care workers in informal conditions of employment. These insecurities play a role in the development of home care agencies. While some grey areas are seen as problematic for some agencies, others have profited from precisely these grey areas; the market is characterised by an uneven development of home care businesses, with small and slowly growing agencies that may cease their placement activities temporarily at one end and larger, quickly growing agencies at the other.

The regulation of the agencies matters, as the different working arrangements for care workers arise from the different forms of agencies: whether they are employment or placement agencies, and whether they operate from Switzerland or from abroad.

The mechanisms that monitor working conditions differ, as does the chance to claim rights in cases of violation of labour regulations, depending on whether the employer is a care agency or a private household. Theoretically, the least difficult case seems to be with placement agencies that are subject to collective labour agreements. The care worker who, with the help of the Respekt group, could prove she had worked overtime and was unpaid for on-call time, and so successfully claimed reimbursement, showed this. In contrast, in violations of labour regulations that occur with employment agencies, it is the private households that have to be sued individually. Since private households are not subject to labour law, it is clear that claims to rights are more difficult in such cases. This is especially so when private households do not comply with regulations for formal employment and employment agencies do not feel responsible for ensuring their compliance, sometimes resulting in difficult situations for care workers, as we will see in a later chapter. Given these considerations, it matters whether the SECO pursues a strict policy in asking employment agencies to recategorise themselves as placement agencies, examining their working contract samples, and pursuing agencies operating from abroad.

Finally, we found that many placement agencies send care workers back and forth between recruitment country and workplace at regular intervals, except for Yvonne's agency, which arranges for care workers to stay for at least half a year in Switzerland. While some employment agencies organise similar rotation schedules to those of placement agencies, others make more use of short-term stays, sometimes even placing care workers for up to only three months a year. All these arrangements are facilitated by the interplay between business strategies and migration regulations; these result in new forms of working arrangements in Switzerland, care workers frequently travelling from home to workplace and back, and short-term employment contracts and ever-changing care workers for care recipients. All of these require relatively high mobility on the part of the care workers.

5   Finding a job in live-in care

Mattea, owner of a care agency in Slovakia, and I were sitting in the front seats of her car, and Sara and Emilia were in the back. It was 7 a.m., and we were in Zurich on our way to the women's very first placements as care workers. Sara liked Zurich. She had lived in Switzerland before. After graduating in psychology in Slovakia, she had trained as a care assistant in Winterthur, a small town near Zurich. During that time, she had picked up Swiss German and was now excited to be back in Switzerland. Later, in an interview on Skype, Sara told me that she had only applied around ten days prior to her placement as a care worker. She had been temporarily unemployed when she came across a job post on Facebook. 'I called', she said,

> and she [the recruiter] said, send your CV and we'll see. Everything went very fast. Within two days she called me to say, yes, so we have a man, he needs help. He works part time. And so I said, yes okay, I'll try this. I didn't expect that. I sort of did it for fun.

The recruiter subsequently sent her a contract, which she signed and sent back within two days. 'I don't really remember. But it was very fast. That woman [the recruiter] was nice. The way she talked was nice', Sara recalled. She was surprised that she was placed so fast.

While Sara was placed through Mattea's agency, another care worker, Krisztina, came to work in Switzerland with the help of a Hungarian transport firm called Interland. She was introduced to Interland through her friend Eva. The firm, which she contrasts with private individual agents as the only 'company-like' agency, also advertises online. When she called the firm, she immediately mentioned that it was her friend who had recommended Interland to her. The person at the other end of the phone was already expecting her call, as Eva had informed both parties about her recommendation. Similar to Sara's experience, Krisztina was placed in a position quite quickly. Within one week of contacting Interland, she was with a family in Switzerland.

Both Sara and Krisztina considered themselves lucky that they were placed so fast. Many other would-be care workers waited a long time and found it difficult to secure live-in care work in Switzerland. Why are some care workers placed very quickly while others wait longer? In this chapter, I first examine a range of factors that influence the uneven mobility of live-in care workers, their 'motilities', or capacities for mobility, and present the contexts that delimit or enable migration for live-in care work (Kaufmann 2002, 37; Kaufmann, Bergman, & Joye 2004, see Chapter 4). This shows that their move into live-in care work is a multi-layered decision process and not, in most cases, a straightforward one. Subsequently, I explain how care workers learn to look for a job in live-in care work and how information on job opportunities circulates. I also describe care workers' perceptions that leaving their villages for live-in care work is nothing unusual and yet affects and changes the places they come from.

Different starting points for live-in care work

While Sara and Krisztina's journeys into live-in placements happened relatively fast, Ilona's case is entirely different. Having worked at the post office in a Hungarian village since she was 18 years old, she had been toying with the idea of going abroad as a care worker for a long time. 'The children have grown up, see, and you'd like to fly off a bit, and you'd like to do something else' she said, 'not this monotonous work where you have to leave every morning and work until evening. I wanted to try it for a little bit'. Ilona saw live-in care work as a chance for new experiences abroad. Before starting her first placement, however, she wanted to be ready and started preparing by attending German class in the village. This was organised by the agency that recruited her. Many of the care workers who registered with the agency later than Ilona had already gone to placements, but Ilona had not. She did not feel ready yet, and she was not in a rush to start her first placement because she had a stable job and considered herself lucky for that. Hence, she took time to prepare carefully for her first placement. By comparing herself and her situation with other care workers that she knew, Ilona was very much aware that the women in her region left for care work from very different starting positions. Her own situation presented four salient aspects.

Firstly, she knew that her German skills were not as good as other care workers' yet, so she wanted to improve her language skills. Thus, she acquired motility by learning the language. In contrast, her cousin was further along in this process; she had already learned German. Thus, when the agency first came to their village, 'her language knowledge was very suitable', she explained. Moreover, Ilona mentioned a woman that had only recently joined the agency but had long-term experience as a care worker in private placements. That woman had been commuting between her placement as a care worker in Germany and her stints in Switzerland for several years. In contrast, Ilona did not have any experience of live-in care work and thus had less motility.

Secondly, Ilona's plan to work abroad was not only for adventurous reasons; she also took financial issues into account. She had loans from her bank and wanted to pay them back. However, she was not in immediate financial distress because she had been employed in a stable job for many years. In comparison to her cousin, who in Ilona's words was 'more forced into it financially', she was not dependent on working as a live-in care worker to secure her income; she considered it an option to improve her financial situation.

Thirdly, live-in care work seems to be easier for those who have fewer ties, and in her words, 'constraints here at home', such as family. 'She doesn't have a husband, she's single. And she can arrange to go next to her job', Ilona explained about her cousin. In contrast, Ilona has three children and a husband, from whom she had not wanted to be away when the children were younger. Now that her two older sons were attending university while the youngest one was still in school, her idea of working as live-in care worker had become more feasible.

The only downside Ilona saw in live-in care work was that the agency only enables care workers to work for three months a year, in three one-month stints, leaving the care workers without any income for the rest of the year. This was not compatible with her job at the post office, and she would have to give that job up. However, she was not ready to resign from her job before gaining experience first on at least one placement. Therefore, going on a live-in placement depended on whether would-be care workers were able to reconcile placements abroad with their own activities and jobs at home as well as on the length and stability of the care work arrangement itself.

These four aspects – language skills and care-work experience, financial situations, family situations, and employment situations at home and length of placements – can thus be seen to have played a role in some care workers' decisions to go on placement. They thus form a useful basis from which to compare Sara and Krisztina's starting points. Sara had an education in psychology, was trained as a care assistant, and spoke fluent Swiss German when she applied for her job as a live-in care worker. Sara was not married, did not have children, and was unemployed at the time. Consequently, as the placement was for just two months, Sara felt able to more or less spontaneously venture on her first placement. Krisztina, who was placed as quickly as Sara, had extensive experience in care work in Germany before she started to work in Switzerland and so felt confident about her German skills. She is married, but her children have grown up and left the family home. Her husband was already used to her working abroad, so it was not a problem for her to leave within one week. To sum up, Sara and Krisztina seem to have higher motilities to venture into live-in care work abroad than Ilona. In comparison to her cousin and to Sara and Krisztina, Ilona's financial stability, employment in the village, and ties to her family can be understood as immobilities, or as 'moorings' (Urry 2003, 138), which compromise her motility.

The search for a job in live-in care work

Having demonstrated that care workers have different starting points from which they venture into live-in care work, the question remains: How do they find that work? Hanson and Pratt (1995) claim that women in female-dominated work tend to find jobs through other women in their social networks rather than through formal channels. This is not surprising in view of the advantages that finding a job through informal contacts can bring. Informal recruitment can provide an insider's view about what to expect and information about the norms of care work, salary, and other aspects of the job (England & Dyck 2012, 4). However, this then raises another question: What is the role of intermediaries and employers in the job-finding process? The next sections discuss care workers' job searches and the connections that are crucial in these processes.

Learning to search for live-in care work and support on the Internet

Whether and how fast would-be care workers find employment also depends on who they know that can facilitate mobility and advise them how to find a job. In live-in care work, the job search not involves only gaining access to new employment but also migrating to a different country. According to Papadopoulos and Tsianos, migration is a process that depends on people and things and that is inevitably managed by reciprocal relations (Papadopoulos & Tsianos 2013, 190). Claiming that 'reciprocity between migrants means the multiplication of access to mobility for others', these authors introduce the term 'mobile commons' (Papadopoulos & Tsianos 2013, 190). It involves migrants and other actors sharing knowledge, cooperating in affective relations, and caring for and supporting each other (2013, 179). Accordingly, the mobile commons exist because migrants 'cultivate, generate and regenerate the contents, practices and affects that facilitate the movements of mobile people' (Papadopoulos & Tsianos 2013, 191). Platforms and media for the collection and distribution of knowledge play a crucial role in the mobile commons, as do informal economies that support migration (Papadopoulos & Tsianos 2013, 192). The authors understand the mobile commons not just as a reaction to the discourses and practices of migration control but also as developing their own dynamics. While I find the concept of mobile commons useful for foregrounding care workers' agency and their search for care work not merely as reactions to economic forces and governmental migration control, it does not take into account the dynamics of inequality in access, distribution and sharing of knowledge, and support, and hence neglects the differing motilities of migrants in the mobile commons. In this section, I present the stories of Berta and Marina to show how care workers learn to search for live-in care work, share knowledge, and support each other on the ground, and how the outcomes of their individual searches differ.

Berta used to work as an educator and teacher in Hungary. Due to health issues, she was advised by her doctor to give up her job. When she travelled to Canada to visit her boyfriend there, she came to know about a Hungarian family offering informal employment. They took her on as a care worker, and she worked there for three months. After breaking up with her boyfriend, Berta came back to Europe and started to look for live-in care work. Later she came to Switzerland as a care worker through an agency in Hungary that places care workers in both formal and informal jobs. Her friend showed her one of the company's advertisements. Berta called the agency and then drove down to see them. As she recalled her experience of applying to the agency, she explained that

> they checked my language proficiency, I had to write a very short, not even CV, but an introduction, of what I want, how, what is my experience, and so on. And they also checked, how should I put it, basic intelligence, which is necessary for this job. Because we go to places where we have to talk. If they say Mozart, I shouldn't think they're talking about chocolate.

After around three weeks, the agency placed Berta in a Swiss household. 'When I started, I was enthusiastic, like why not? I don't have anybody. Only my son, who is studying, so there wasn't anything to go back to', she said. However, after a while her attitude changed: 'I thought it would never end! It was exhausting!' The care recipient actually wanted Berta to stay for half a year on her first stint, but Berta wanted to see her son and take care of her house, so she agreed to stay for three months. In that household, she met an outpatient carer (a *Spitex* employee) who directed her to another family who offered a better salary. Subsequently, she changed from the agency to being directly employed.

When we talked to Berta about different care agencies and private intermediaries, she immediately went on Facebook to recall some of the contacts she knew: 'There's one here in Kaposvár. And this lady for instance, is the one I wanted to suggest to you. In Szentbalázs, that's here. It's also a care group, a closed group [on Facebook]'. In the course of the next 15 minutes, Berta browsed on Facebook, the social media and social networking service, trying to recall the names and groups that she had come across and used when she was still looking for a job:

> Here, there's a man from one group; he writes male care giver is needed, with immediate start, for a job in Germany. We as private people, the members in the group, can advertise such things. There are others that do the transport, they also advertise, because not everyone can reach [their workplace]. Driver service. *Utazz velünk* [travel with us]. The jobs that we advertise here, I don't mention them elsewhere. Unless you put up a profile picture, they won't talk to you.

Before she started working as a care worker, she did not really use Facebook. Later, she was even able to find a 'switch partner', another care worker to replace her in Switzerland while she was back home, through Facebook:

> I helped this Erika without knowing who or what she is. I just advertised in this closed group that I was looking for someone as a relief, the work is about this and that, and she replied. But I didn't ask for any money, it didn't even cross my mind.

In sum, Berta gathered her first experience as a care worker through private contacts, which seems to have raised her motility, which then helped her when she applied to a care agency. While working with the agency, she gained the opportunity to change to being directly employed through another private contact. Later, she even became an intermediary herself through Facebook.

Marina, a good friend of Berta's since childhood, also actively used Facebook in relation to care work, initially to find employment and later also to support other care workers. Feeling that her 'financial situation in Hungary and other circumstances made it simply unavoidable' to try and find work abroad, she quit her job as a kindergarten teacher in Hungary, where she had been working for 33 years. She raised her son alone as a single mother. He is a professional athlete yet depended on her support for his career. To support him, she felt she needed to find work abroad. He was 18 when she found her first job abroad, as a kitchen porter in Austria. The working conditions were hard, and after experiencing abuse from her employer there, she quit and found work as a nanny in Germany. However, after four months she did not receive the salary that she had agreed with her employer (1,300 Euros) and did not receive her days off as agreed (two days off in two weeks). So, she started looking for another job. Having worked as a nanny gave her the idea and confidence to find a job in live-in care for the elderly.

When she started to look more closely for employment as a live-in carer, she first sought acquaintances with experience in care work and asked them how they had found work. That is how she came to know about individuals that act as private intermediaries. 'They didn't ask for a CV. One, when I phoned to ask, sent me a form that I had to fill in with my personal details', she recalled. However, the agents she contacted were offering to place her for a fee. 'I didn't have a penny, let alone enough to pay a placement fee'. On one occasion, Marina paid an agent a placement fee of 15,000 forints (around CHF 50), which, in the end, did not lead to employment. After her encounter with the application forms that the agent had sent her, she started to create a CV with the help of a dictionary and an online translating machine. The questions on the form helped her decide what to put in a CV, which she later sent to larger agencies.

Marina started to use Facebook in 2008, when an old friend from Finland invited her to the site. She had mainly used it to keep contact and exchange messages with friends and acquaintances that were abroad. 'Later this changed, you know. So, I got this tip-off that you can ask this person about jobs. The first private contact, if I remember right. First, they always check language knowledge. You're always tested on the phone', Marina said to explain how she first started using Facebook to look for care work. She tried several channels at the same time, as she wanted to start working as soon as possible.

> I stopped when it turned out that these people often ask for a placement fee. And from then on, I looked for formal placements. Afterwards I was using the Internet pretty seriously; I learnt how to find everything on Google. This is how I found the websites, and I applied to these.

However, for her very first placement in live-in care, it was an acquaintance that told her about the informal opening.

I said yes, because I'd been waiting for a long time and I wanted to work. And this is how I got to Germany; that was my first elderly care work. I could hardly wait for the first opportunity, so I jumped at it.

After her first placement, Marina kept looking for other opportunities until she found work in Germany with an agency called Hausengel. With Hausengel, she worked in informal arrangements, but she later switched to being directly employed, where she earned less but felt more secure.

Overall, Marina learned over time how to use Google, how to research opportunities, and how to distinguish scams from genuine openings online. While showing me and my research collaborator Katalin a post on Facebook, she explained:

It says they're offering a salary of 1,500 Euros net. So someone who has experience knows right away that this is a scam. You can tell because, on the one hand, there are no details, who, where, under what conditions, and so on. The most striking is if they write a net salary. A foreign company never lists net salary, only gross. And [they don't list] an hourly wage.

Moreover, Marina says, a foreign company would not advertise with a Hungarian care agency but work with a recruitment company that would publish an advertisement. These recruitment companies have official websites and tax numbers. While Katalin and I would not have been able to tell whether that specific post was a genuine opening or not, Marina had learned how to distinguish between advertisements and posts on Facebook and felt confident about telling whether they were trustworthy or dubious.

It seems that care workers exchange their experience of scams online and in Facebook groups. Marina heard about stories in which brokers bring care workers to their workplaces, 'leave them on the street, make them pay the travel costs, and just leave', and stories about care workers that were placed but never received salary. Having collected knowledge and experience, Marina decided to actively support other care workers on Facebook:

I decided, since I'm in elderly care and confined [to the house], I don't have a community, and I've always considered myself a social person, I decided to create a group where I can help those acquaintances of mine who've asked me for help. . . . You can see everything in the group, and we learn from each other. There aren't so many of us, around 150. Almost half of us work abroad. We form very good human relationships in the group. I am trying to pass on my experience. For example, I know you shouldn't pay, and I repeat that like a parrot. Don't pay. It's better to make the effort and write a CV, even if it's hard, but she should do it, create a cover letter. She can get samples, help for that. So if an advertisement appears on a foreign website, and she's sure about it [that it is a serious company, she can apply] and not be taken for a ride by adverts from Hungarian intermediaries.

The care workers upload documents and examples of their own experience into the Facebook groups. Hence, anyone that is in these groups has access to this knowledge and can actively contribute to it.

In 2013, Marina and Berta planned to work together in the same household in Switzerland as switch partners. Berta had found a contact, and knowing that Marina was looking for a job, she asked Marina if she was interested. Since Marina's German skills are better than Berta's, Berta asked Marina to write the CVs and the letter to the family. However, for reasons that neither Marina nor Berta could explain, only Berta ended up getting the job. Although Marina's German skills are better than Berta's, so she would have better chances of working as a live-in care worker than Berta, this did not matter; in the end, she depended on Berta, who acted as gatekeeper in the situation. To conclude, not only do the motilities of Berta and Marina consist of very different components, but the outcomes of their job searches also differ. In her current position in Germany, Marina works permanently without a switch partner. While Berta earns 1,650 Euros, Marina earns half as much: 800 Euros per month. Even two years after our first interview, Marina was still searching for opportunities to work in Switzerland. Although both are in some of the same circles and groups in which information circulates, their journey from one household to another in live-in care work led Berta to work in Switzerland and Marina to work in Germany.

Differences between finding work in Switzerland, Germany, and Austria

The difficulty of finding work in Switzerland was also an issue raised at the group meetings of Respekt. When I asked whether it was difficult to find work, the answer was a clear yes. The impression I had from my visits was that it seems to be especially difficult to find direct employment. The general advice the group members give to each was that it would be better to find employment privately and directly instead of going through a home care agency, to avoid placement costs. However, this also limits the opportunities to find work.

Care workers look not just for any employment but also for more sustainable long-term employment that is especially difficult to find. Anna, who I visited in a household where she was employed as well as in her home village in Hungary, was working in two households in Switzerland. She worked one month in one place and the next month in the other and went home for a couple of days in between. Sometimes she would only spend two days at home with her family. She was exhausted, had health issues, and was dealing with conflicts in one of the households. As a result, she had started to look for other opportunities. Over several months, she repeatedly told me how difficult it was to find a new job that would improve her situation without paying a lower salary than her current placements.

Generally, more knowledge and experience circulate about live-in care work in Germany and in Austria, and the overall perception of the care workers interviewed in Hungary was that fewer care workers go to Switzerland. This is because an informal live-in care market, characterised by informal and word-of-mouth networking, has existed between the EU-8 accession states and Germany and Austria for longer than with Switzerland. Moreover, transport businesses and care agencies placing live-in care workers developed much earlier in Germany and Austria than in Switzerland (Krawietz 2014, 13; Österle, Hasl, & Bauer 2013). As care agent Dora puts it:

> Unfortunately, the financial situation in Slovakia, pay, is low. So plenty of care workers are interested in live-in care work in Switzerland. So that's not a difficult situation for me, because many have worked as care workers in Austria and in Germany, and now they're interested in Switzerland too. From a Slovakian perspective, I would say that Switzerland is a recent extension. Before, we could only work in Austria, and then in Germany too, and now in Switzerland as well. The market has expanded.

Thus, access to live-in care employment is easier in Germany and in Austria than in Switzerland.

Circulation of knowledge through word of mouth

Care workers share knowledge not only on online social media platforms but also through direct word of mouth. During her work at the post office, Ilona talks to the villagers, so she also hears about their experience in their everyday lives. 'I listen and it's so much. We talk to each other [about living in other countries] and it's inspiring. I regret that I didn't learn languages', she stated. The post office is also where she came to know Irina, the German teacher, and Krisztina, who helped Pascal start his agency. Although Ilona had no experience of live-in care work yet, she had already passed on recommendations to the agency:

> I know a lot of women. I even recommended ladies from another village. The care recipient of this woman died. She'd been working as a care worker abroad for years, she speaks German, she's got experience, and Pascal was happy with her. At the post office, I tell anyone I can. I talk to people.

Ilona's example shows how experience and information about care work opportunities are transferred from person to person. The circulation of knowledge here is embedded in the activities of Ilona's everyday life and so is spread in a passive rather than active way.

Berta and Marina found work mostly through private contacts and agencies that place care workers in informal employment. Another way to find live-in care work is with the help of private transport firms in the recruitment countries. 'You must have heard about them', one of the care workers told us during our group conversation in Kicsifalu, where Katalin and I joined a group of care workers who had met in a café.

> The driver finds someone, or asks someone, if you know about something. They've got networks in Germany. They've got contacts with families and acquaintances. The drivers get a lot of addresses this way, and you can look for a place through the drivers.

When I asked if these drivers are from a company, another care worker answered:

> No they're private drivers. The news spreads by word of mouth. The women sit in the bus and they ask each other if they know someone from this place or that place. And the driver, if he's smart enough, he writes that down immediately. This is how he starts it; yes, I can take you, but it means that you have to travel with me.

While some care workers share their knowledge about live-in care work and the experiences they have had with various care agencies and in different places, others prefer to keep such information to themselves or only talk about it with care workers in similar working arrangements. For example, Anna told me:

> I don't talk to some of the other care workers in our village, because some of our salaries are better than those who work with an agency or who work in Germany. One time, a woman that I know called me and told me about a job in another household. I turned down the offer, because she doesn't know how much I earn and I think her offer would have been less.

When I visited Anna in Kicsifalu, she had arranged an interview for me with her cousin. Before the cousin arrived, Anna told me that she herself would rather not participate in the interview. Her cousin apparently did not know how much Anna earned in her live-in care job in Switzerland, and she suspected that her cousin earned less. By not sharing certain information about her live-in care work, Anna wanted to avoid embarrassment and conflicts that might arise from her better working conditions. Anna's example shows how access to and sharing of knowledge is very much embedded in social relations and everyday social norms.

The spreading of knowledge has its own dynamic, so some can access knowledge and opportunities better than others. As Krisztina puts it: 'Actually, very much depends on the contacts here. If you're already in the circle, we keep and pass on contacts among ourselves, but women are able to charge each other [charge money for information]'. Not only can access to knowledge and job opportunities be reserved for the members of social groups, but opportunities also depend on whether care workers have the financial means and are willing to pay for access to information. Finally, according to Krisztina, 'those with bad references don't get placed again'. Hence, circulation of information can also have negative outcomes for live-in care workers.

Going abroad for care work as nothing unusual

The circulation of information on live-in care work is also connected to how commonly care workers migrate from particular places. Elrick (2008) and Elrick and Lewandowska (2008) have shown that networks play an important role in facilitating migration where specific migration networks are long established. Moreover, drawing on Massey et al. (1994) and Horváth (2008), Elrick (2008, 1505) argues that 'cultures of migration' have emerged in which values and cultural perceptions are changed over time by the exchange of cultural symbols, information, and technologies attached to migration. When I visited Kicsifalu, I was struck by how going abroad for work was treated as common. It seems that ever more people have left the village within the last decade. 'Not long ago, there were 4500 of us. [Now there are] just 3574 people', the mayor of Kicsifalu said. In the course of a half-hour conversation, Anna and the mayor told Katalin and me stories about numerous people, many of them care workers who left for work abroad, and they named whole families that had even left for good:

> Every young person leaves, that's for sure. Their families don't prevent them from leaving. They can't even say anything; they themselves go abroad to work and so give a bad example. The Szabos, a family with three children, they moved away. Just think about it! In Kicsifalu, a town with 3500 inhabitants, a family with three children leaves. It's a loss to the kindergarten, to the school. . . . And this is just one example! . . . Children are missing from kindergarten, from school. There will be fewer and fewer children, so fewer and fewer teachers are needed, they become unemployed.

When I asked how many women in the village did live-in care work, he answered: 'I often think about that, but it's hard to estimate. Around 150? It might not be that much. It could be a lot of fuss about nothing [*viel Rauch um nichts machen*]. But we can easily and quickly name at least 30 to 40'. 'Yes, just the ones that we know about. Because many leave without notice', Anna added.

According to the mayor, the cause of increasing numbers of people going abroad to work was economic changes in the area arising from the closure of agricultural cooperatives (*termelőszövetkezet*) and other employment opportunities. He recalls that the two main cooperatives in the area used to employ around 1,700 to 1,800 people, and that now, due to privatisation and other changes, they only provided work for around 90 employees. 'Everyone used to have a job. They even came from out of town. People lived in security', Anna commented. The mayor agreed:

> We used to hire a lot of women, even from other villages. 400 women worked in the shoe factory, and most of them were unskilled workers. More and more of those women go abroad to work, and to my surprise, even those without any language knowledge go. . . . [However,] care work is a relatively good source of income for the locals. But it's not good for Kicsifalu. Families fall apart.

Anna and the mayor both stressed that care workers work abroad primarily out of economic necessity. They implied that the large numbers of people leaving would have serious consequences for the village. For example, the mayor and Anna talked at length about their personal experiences with care workers' families facing tensions and instability because of, to use the mayor's word, the 'reversed' gender roles of live-in care workers' families. They suggested that an ageing village would be left with elderly in need of care.

Conclusion: altruistic and profit-oriented networks entangled

This chapter explained how care workers find work in private households in Switzerland and what role care agencies play in this process. I showed that care agencies can serve as entry points for care workers to a placement in Switzerland, which in turn can lead to further contacts for direct employment. However, care workers have diverse starting points and access to live-in care employment. Those with better German skills and experience in live-in care work and without care dependents such as children at home seem to have higher motilities and are more likely to be hired by care agents. Moreover, the financial situation of potential care workers and whether they have stable work at home affects the decision. It can be difficult for care workers to reconcile live-in care work with their own activities and jobs at home, especially when working with care agencies that place care workers for short stints and do not guarantee a stable income. Hence, care agencies' working arrangements play a role in care workers' access to live-in care work, as does whether care agents take the stability of care workers' incomes into consideration.

In the job-finding process, care workers learn to search for care-work opportunities and to distinguish between serious offers and scams. Overall, a great deal of information circulates on live-in care work: Care workers share information on live-in care in their local communities, pass on recommendations about brokers, and share knowledge on how to apply to live-in care agencies. Social networks and the circulation of employment opportunities in the home communities of care workers constitute an essential part of care workers' migration infrastructure by providing care workers access to employment. The circulation of information is both embedded in the everyday lives of care workers, potential care workers, and agents and happens both in online spaces and by direct word of mouth. How prospective care workers access these networks and this information is characterised by their social skills and follows social norms for the creation and maintenance of social relationships. Care workers actively support each other in reciprocal relationships, sometimes in exchange for monetary payment. Or they decide not to pass on information to avoid social conflicts and tensions, such as when Anna decided not to tell her cousin about her salary. Moreover, it was important for some care workers to maintain access to agencies they had worked with even after they had found direct employment through private contacts. Krisztina even paid an agency a fee for a journey she had not taken to maintain a good relationship with them, despite feeling that the agency should not have demanded the payment.

Much of the information sharing and support is provided by talking, and word-of-mouth recruitment and online on social media platforms. Care workers such as Marina and Berta have learned to navigate the online space and distinguish serious care work offers from scams. They actively exchange information in Facebook groups on working conditions, how to find employment, and how to apply to care agencies. They can connect to their support networks and contact informal brokers and home care agencies from both the living rooms of their workplaces and their homes in Hungary. In this sense, the Internet has emerged as an essential tool that facilitates care workers' communication and access to live-in care work across national boundaries, thus leading to the creation of new transnational ties.

However, information about opportunities for work is not equally distributed in what Papadopoulos and Tsianos (2013, 179) call the 'mobile commons' and can sometimes only be accessed by paying for it. In this context, the interviews show that a distinction between altruistic social networks and profit-oriented agents dissolves on the ground and that profit and trust in private networks are very much entangled. Having showed how women find work as live-in care workers, the next chapter addresses the question of how agencies recruit care workers.

6 Recruiting care workers

Every two weeks, Andrea, a recruiter, would take a flight to her recruitment site in Germany to conduct face-to-face interviews with the prospective applicants that had passed the pre-screening process. In the morning, she would conduct six or seven interviews, and in the afternoon, she would organise the first orientation session for those who had been selected in a previous recruitment round. During the orientation, the new care workers would receive their staff badges, be trained in the general principles of the agency, and watch a film about the company. The agency had one of the most elaborate and strict recruitment processes of all the care agencies in this sample. In the pre-screening process, the potential care workers received a questionnaire to fill out and a handbook informing them about the agency, their working conditions, the prerequisites for working with the agency, and administrative procedures for a placement in Switzerland. The applicants then submitted the form along with six references, three from private individuals and three related to professional experience, a criminal record statement, and an excerpt from the debt collection register. Those who were deemed suitable were invited for an interview in Germany. After setting up and managing the whole recruitment process by herself for ten months, Andrea was given the help of an assistant by the agency so that she could delegate the pre-screening process. While the pre-screening process can be conducted through communication technologies, the agency found it important to include final face-to-face interviews before actually hiring a care worker, which is why Andrea still travelled regularly to Germany.

Having presented how care agents started and developed their home care agencies in Chapter 3, I continue here to provide insight into their business practices by showing how care agencies recruit migrant live-in care workers. It complements the previous chapter in understanding the roles of agencies in care workers' journeys in finding a job. I first show how care agencies access potential care workers. Care workers can be recruited both directly and indirectly; in the latter case, recruitment is outsourced to a third party. Subsequently, I describe how care agents create the flexibility required for their live-in care services by creating pools of disposable care workers. The third subchapter is dedicated to how care agents select care workers and create the subject of a warm-hearted, selfless older migrant woman from Eastern Europe. I conclude by arguing that these recruitment practices contribute to the creation of a gendered migration channel from Eastern European countries to Swiss households.

Setting up transnational recruitment infrastructures

Home care agencies that recruit and select care workers advertise care work opportunities directly on their websites, on social media platforms, and in other job portals. For example, one of the agencies places an advertisement in an Eastern German newspaper twice a week. 'We tried different things, and we found that this is the best way for us. This is how most people contact us', Andrea explained. The agency only recruits from Eastern Germany, emphasising German language skills as one of the key criteria in their recruitment. Andrea co-developed the whole recruitment process and is in charge of the allocation of care workers from the main office to all 21 offices in Switzerland. She grew up in Germany and came to know about the agency because her mother had already worked as a care worker for the agency. While working for a hotel as a marketing associate, she happened to meet the agency's managing director, who was staying at the hotel, and he offered her a job with the agency in Switzerland, first in marketing, later in recruitment. Like all the other interviewed agencies, Andrea recruits migrant care workers from countries with lower wages east of Switzerland. Hence, unlike local recruitment, job applicants cannot be invited to a job interview at the place of employment but have to be assessed through either communication technologies or other kinds of infrastructure to facilitate these transnational relations. Some of the agents, including Andrea, travel to the places of recruitment themselves to recruit care workers on site.

Travelling to recruitment countries to find care workers

Owners of smaller care agencies, such as Pascal and Dora, also travelled to their recruitment locations from time to time. Dora, who grew up in Slovakia, combined recruitment with her private travel to Slovakia. As she had worked as a live-in care worker in Austria before she came to Switzerland, she believed that she had an advantage over other home care agencies. In addition to her existing private social network in Slovakia, she emphasised the importance of having good working contracts and working conditions, which she tried to ensure for the care workers that she placed, so as to attract good care workers. In contrast, Pascal's trips to Hungary had the sole purpose of finding potential care workers. He started his whole business and recruitment based on his relation with Krisztina, the care worker that he had employed to care for his mother. After talking about the economic situation in and around the village that Krisztina comes from, Pascal developed the idea of recruiting care workers from that village. She told him that there was not much work, salaries could be as low as 300 Euros per month, and that there were a couple of people who could speak German. That was reason enough for Pascal to visit the village and see for himself. Since he does not speak Hungarian, a translator had to be organised. That was not difficult; Irina, a

German teacher in the village, was available to help. And as word of Pascal's visit spread quickly, around 30 women gathered to hear what Pascal had to say. With the help of Irina and Krisztina, Pascal decided to set up a German class to prepare potential care workers for their placements, and Irina became the coordinator for the care workers.

When Katalin, my research collaborator, and I visited this village, we found Irina's role as a coordinator to be quite significant. She gave us the impression that she was very much in charge, even before our arrival, and showed us clearly that she was responsible for whom we could talk to, when we could talk to them, and under which circumstances. Irina had organised a tour of the local history museum and interviews with five care workers for us, but initially provided a time slot of only two hours. She also arranged a dinner with the president of the German minority association. I would have liked to stay longer in the village and take more time over the interviews. However, we did not appear to have a say in this, and we were grateful that we were able to visit at all. Overall, I left the village with the impression that Irina plays a significant role in the village community as a gatekeeper for would-be care workers. Hence, in addition to travelling to the recruitment village himself, Pascal the agent started a close collaboration with one of the locals in the village.

Care agent Daniel decided to set up his own recruitment agencies in Poland because his wife, who also works for the agency, is from Poland, and the couple frequently stays there. For him, face-to-face interviews are 'a must' for recruitment, and since his agency only has recruitment firms in Poland, he recruited from Poland and not from other countries. Mattea, who is Swiss, even decided to move to Slovakia and base her agency there. She undertook occasional trips to Switzerland both for private and business purposes. However, although her agency is located in the recruitment country, most of their recruitment was based on e-mails, Facebook advertising, and phone calls. Face-to-face interviews were only conducted if it was convenient for care workers living nearby. For the recruitment process, Mattea hired two women, Leila and Nina. Leila had experience in the field of live-in care work before she applied for her job as a recruiter. She had been commuting between Austria and Slovakia every two weeks for seven years before Mattea offered her the job at the agency. Nina was working as a waitress at a restaurant where Mattea often had lunch. Mattea knew that Nina had worked in Switzerland before as a waitress and offered her a job at the agency. 'They are two alpha-type leaders', as Mattea described them; 'we've arranged leadership training for them'. The two recruiters worked closely with the two sales employees, both of whom had experience working as live-in care workers in Austria. As soon as they received an expression of interest from a care recipient, the recruiters posted a short job advertisement on their Facebook homepage with basic information about the location, duration of placement, and the care recipient's health status. Usually, the agency was able to fill the position within one to three days. Subsequently, the recruiters summarised the data of the shortlisted care workers and sent a selection of three CVs to the potential care recipients. Direct recruitment involves a mix between communication technologies and travelling to or, in

Mattea's case, even moving to recruitment countries to facilitate the transnational relations required for the recruitment of migrant care workers.

Working with recruitment agencies

Before Mattea decided to begin recruiting, she had collaborated with small recruitment agencies in Slovakia. Collaboration with recruitment firms can be convenient for care agencies who do not want to recruit themselves or who are just beginning their own recruitment processes. These recruitment firms are usually located in the recruitment countries and can be either smaller firms specialised in home care or larger agencies that recruit workers in a range of sectors. The way these firms function seems to be similar to call centres. Mattea even sent a colleague to work in one of the big recruitment firms for two months to gather information about their recruitment system, or, as she phrased it, 'to conduct industrial espionage'. The big recruitment firm employed up to 60 call-centre agents, whose main task was to call job-seeking applicants and compile lists of workers. The recruiters scanned advertisements for care workers and profiles on job-seeking webpages and other sources. They then called and registered their data. According to Mattea, the call-centre agents were obliged to call a certain number of people every day. The data was then offered to firms seeking to hire employees. Thus, recruitment agencies may play a crucial role in facilitating recruitment for care agencies without direct access to care workers.

Anthony, owner of a placement agency, does not even look at unsolicited applications but forwards them directly to his partner agency in Slovakia or leaves them unanswered. For him, it is easier to outsource the whole recruitment process so that he does not have to deal with finding care workers, screenings, assessments, and selection and qualification issues. Not only are the costs lower for him when working with a recruitment partner in Slovakia than recruiting in Switzerland, but he believes that he can also avoid all the responsibilities that come with recruiting an employee. 'I only place them; he [the recruitment partner] is responsible for quality', he says. 'I never see the women personally, only their pictures'. He does not know how his partner recruits and how large their pool of care workers is, but he assumes that 'it should be big enough', as they also work with partners in Germany and in Austria. He even delegates the final selection to the family. His partner sends two to four suggestions with pictures of potential care workers, which the care agent then forwards to the family. What is interesting here is that Anthony never has direct contact with the care workers, and his recruitment partner is never in contact with the care recipients or their family members.

As can be seen, transnational recruitment here functions through a chain of intermediaries. Each intermediary operates locally, and information is transferred by e-mail, phone calls, and webpages. The only mobile human subjects that travel across borders in this arrangement are the care workers. If the recruiters and the care workers do not meet directly, the only face-to-face contact is between the care recipient and the care workers at the end of the placement; every other step occurs through information and communication technologies.

While outsourcing recruitment to agencies in lower-waged countries seems to work well for Anthony and some other care agents, Mattea is much more sceptical about working with recruitment firms: 'So they send 15 CVs and then in the end a completely different person comes, because none of them actually really wanted to work'. Her scepticism is based on her experience with several recruitment firms where placements did not work out:

> Just last month we had to bring two home. One didn't want to work. The other one was constantly complaining. Everything would be better in Slovakia, where she has a big house and a big car, and she didn't have anything in Switzerland. They were both brought in by our old recruitment partner. So we had to say, okay, go home. Enjoy your big house and your big car in Slovakia. Don't go to Switzerland, you knew what was expected. We are not sure whether that recruitment agency informed the women properly. We can be sure that the ones we recruit ourselves are well informed. Just as our client gets a CV, our care workers get a CV of our client. We've had nothing but problems [with recruitment partners]. If there's a problem with a woman, then you have to check with the recruitment agency first. And you pay good money, half or 75%, but in the end, you still have to do all the hard work and deal with the problems. Finally, you can't serve the client in the best way, because there is always an intermediary who wants to have a say too. Like, why does she have to come back, did she really do something wrong? Now we have our own recruitment; it's more work, but in the end, you can satisfy the client much better. Both sides, care workers and Swiss clients.

Hence, Mattea's independence of recruitment partners meant she had more control of who was selected and more flexibility, as she did not have to justify the termination of a placement. However, even though the agency had already set up their own recruitment, Mattea still worked with their previous recruitment partner if her own recruiters were not able to find a suitable care worker.

Activating informal networks

Besides collaboration with recruitment firms and direct recruitment, another means of recruitment is to activate care workers' informal networks and encourage them to recommend other care workers to the agency. Care agencies also often offer their care workers incentives to recruit other migrant women, as a member of the Respekt group reported (AOZ 2013b). For example, Andrea's agency has introduced a bonus programme for their care workers, called 'care worker recruits care worker'. If a care worker recommends a friend or acquaintance and the new care worker is placed twice, the care worker who brought her to the agency receives CHF 100. According to Andrea, about 10% of the care workers find their way to the agency this way. From the care agencies' perspective, word-of-mouth recruiting is a cheap way to build a pool of care workers. Nevertheless, the screening of potential employees to ensure the most appropriate match for the employer seems to be a critical task for care agencies. This shows how formal recruitment and informal networks are very closely intertwined.

Creating flexibility through recruitment

Care agencies often market their live-in care packages by highlighting their screening skills and their skills in finding suitable workers for each household. Moreover, many care agencies offer to change care workers if care recipients feel an arrangement is not working well. The placement of care workers is characterised by their interchangeability; they can be individually matched to specific households and placed relatively quickly. How do home care agencies achieve this disposability and flexibility, which constitutes an integral part of their business model? And how do they maintain it?

Building pools of care workers

Many of the interviewed agencies build pools of care workers. What is striking is the size of some of the pools, especially in relation to the numbers of care recipients. For example, Dominik's care agency had compiled a list of 200 potential care workers and aimed to build a pool of 300 to 400. However, when I asked about the care workers in actual employment, Dominik answered that only around 30 care workers were employed to care for 15 care recipients. 'When we have ten available, we call them, then maybe six don't pick up the phone, we reach four, two are occupied, two are still thinking', he explained to me. During a visit to Seniorenzuhause in Zurich, the managing director reported a pool of approximately 300 care workers. The agency at the time had 7–10 care recipients and therefore only employed 14–20 care workers (AOZ 2013a). Mattea's agency had around 100 care recipients, for whom more than 400 care workers were placed, since the agency's main model was based on short-term placements of up to three months. Similarly, Daniel had built a pool of around 500 to 600 potential care workers within one and a half years and was still in the process of expanding it. However, only 110 to 120 care workers were actually placed for 100 care recipients. Daniel had five recruiters fully employed in three of the agency's own recruitment offices in Poland. When I asked Daniel why they had such a big pool, of 500 to 600 care workers for 100 care recipients, he answered:

> Imagine you only have 150 care workers in a pool and 100 clients. Sometimes you need holiday replacements, or you need someone for a short time, then you need 120 workers at least already. And then the rest of the 30, they don't sit around and wait. They go to other agencies. If they don't get a job in Switzerland, they go to Germany or to England, depending on their qualifications. So you have to have a large pool to be able to react to all demands. And then, as I said, not every customer is the same. I don't want to standardise anyone, so we need as many different care workers as possible that can be

matched to the individual client. If not, you can't do this. If you want quality. And to give the client the service that he really wants and deserves.

For Daniel, flexibility was directly related to what he understood as the quality of his service. His agency was able to organise a care arrangement within 72 hours. 'Flexibility is at the heart of this business', he explained. 'Without flexibility, the business model does not work'. Interestingly, Daniel did not use the same pool for the agency's services in Germany as for Switzerland. According to him, care recipients in Switzerland had different criteria:

> A German client is not a Swiss client. A Swiss client, I'd say, they are all sceptical at the beginning, but finally, a Swiss client hopes to [find], I don't want to say it, almost a family member. Warm-heartedness [*Herzlichkeit*] is very much called for in Switzerland. Personal conversations are important in Switzerland. It is an all-round care package. It's not just cleaning and being on hand for an emergency as in Germany. There, it is not so much about personality; tasks just have to be done. In Switzerland, it's different.

Hence, the building of a pool of care workers is connected to the agency's marketing practices, which seek to frame live-in care as especially caring (more than in Germany). The meaning of 'warm-hearted' care, however, only unfolds in relation to the care recipients, when care workers are regarded as 'almost a family member' and able to interact in conversations with care recipients. The term 'warm-heartedness' or 'sincere warmth' (*Herzlichkeit*) is indeed a central characteristic in care agencies' marketing brochures and in their web presence (see also Pelzelmayer 2016). Hence, Daniel legitimises the need for the large pool of care workers by prioritising the need to satisfy both demands: fast placements and a diverse range of care recipients to match to.

Delegating flexibility to care workers

Pascal had built a pool of 32 care workers within one year, with which he felt he could offer services to at most eight care recipients. He had placed care workers with five care recipients but had set a goal of 15 care recipients in total with a pool of at least 60 potential care workers. The minimum ratio of four care workers for one care recipient here corresponds to his business model of placing a new care worker in one household after a maximum duration of three months for each carer, so that at least four care workers rotate through a household in one year. Although care workers can only work up to three months per year with him, he does not consider how they are able to reconcile such placements with their jobs and activities at home to secure another income than working with his agency:

> I have a lot now, who don't work with me, who work in Germany. I don't care who they work with there. But I don't want, if I call, and I need them, that

they say they are in Germany. Then I say, you have to decide now where you want to work, with me or in Germany.

Hence, Pascal delegates the responsibility of organising a stable income to the individual care workers. They have to figure out for themselves how they can be flexible enough to work with his agency.

The agency that Melvin and Andrea worked for had built a pool of 150 care workers within one year and four months, of whom around ten were taking a break and were hence unavailable for placement. Andrea usually made sure that around 10 to 15 care workers were ready for placements right away, so she could call and ask whether they could take a flight the next day. The rest of the care workers in the pool were either on placement or resting between their placements. Pointing out that often it was not up to her to decide when a placement began, as many of their clients would be released from hospital and need 24-hour care at home right away, she said

> So we always say, the more flexible you are, the better for you. If they tell me that they're only able to start next week, but in the meantime a placement comes up, then it's too bad for them. Then we ask the next person. If they tell me they're sitting on suitcases that are already packed, able to take a plane the next morning, then it's better for them. But it's really up to them themselves. They tell us their availability, we write that down, and then we see.

Similarly to Pascal, Andrea makes clear to the care workers that being flexible would be in their own interest if they wanted to increase their chances of a place-ment. Thus, the responsibility for securing work is again delegated to the indi-vidual care workers.

To conclude, a pool of care workers allows flexibility at several levels. First of all, it allows care agencies to perform some trial-and-error placements. In case the relation between carer and care recipient does not work well, the care workers can be replaced repeatedly until a suitable match is made. Secondly, the pool is also related to whether it is possible to bind the potential care workers to the agency. Care workers are often registered with other agencies and seek work through sev-eral channels at the same time. This is even a reason why Mattea hesitates to offer further training to the care workers, as she thinks that they would change to better jobs if they had the chance. However, it is important to note that the ability to bind care workers to the care agency also depends on the agencies' business strate-gies. It seems that Andrea's agency, with its strict recruitment and a model using two care workers going back and forth in two-week cycles, seems to be able to bind a relatively large number of care workers. In contrast, Mattea's and Pascal's main models both place new care workers every three months. Consequently, they change care workers much more often than some other care agencies with more stable employment arrangements. Daniel employs one care worker perma-nently per household with no 'switch partners'. However, it is possible that his agency places new care workers in one household quite often. Thirdly, and most obviously, the pool is also related to how many care workers a care agency actually needs: how many care recipients they have to match. As mentioned earlier, it is more difficult for agencies, especially the smaller ones, to acquire care recipients than to find care workers. Consequently, the agencies create conditions in which carers have to be mobile and flexible in order to be employed.

Subjectivising care workers as warm-hearted and selfless older migrant women from Eastern Europe

Having explained how care agents gain access to care workers and create flexibility through recruitment, I now turn to the question of how care agents recruit and select suitable care workers. Who is actually recruited as a live-in care worker in Switzerland? Assessing and selecting care workers were major concerns for care agents, especially for recruiters and care agents without previous experience or knowledge of recruitment and/or live-in care work. As Mattea put it: 'How do you classify them by their care skills and whether you can trust them?' The interviewed care agents mentioned a long list of diverse selection criteria. Many named language skills, cooking skills, empathy, flexibility, patience, whether they were smokers or non-smokers, and whether they had a driving license. Some stressed loyalty, devotion to work, whether potential care workers had their 'hearts in the right place', and assessed whether they were 'not too dominant', 'not aggressive', honest, and able to adapt quickly to the households. However, other agents also emphasised that a care worker should be assertive, able to defend herself, and have strong communication skills and 'a strong personality'. Although some of these criteria contradict each other, three aspects of the care agencies' selection process are striking and indicate a degree of coherence across the industry about who is deemed to be suitable as a live-in care worker in Switzerland.

As selfless and warm-hearted

The first key point is related to the recruiters' construction of the 'ideal care worker' as a selfless person with a heart for the elderly. Independently of how strict the recruitment procedure is, whether the agency has introduced multiple screening steps or recruits more loosely through a job interview on the phone, the final selection is often guided by intuition or, as Pascal puts it, based on the feeling of the recruiters. For example, Andrea asks herself during job interviews whether she would let the candidate take care of her own mother. Therefore, even after a prospective care worker has passed the agency's standardised pre-screening process, and even after fulfilling all formal requirements, the final decision of who is selected as a live-in care worker is characterised by individual interactions between applicants and recruiters. The outcome of the selection process depends on how care workers present themselves and how recruiters perceive them.

What is interesting is that many of the recruiters I interviewed tend to prioritise a certain attitude over relevant skills or experience in the health sector. 'Anybody can do it. No qualifications are needed. It's just housework', said Pascal. He stressed that he personally talked to the potential care workers during his visits in Hungary, explaining that

> if someone asks from the start, how much can I earn, how much free time do I have, then I say, well, okay, I won't call you anymore. You have to have a sense for that. You have to, I don't know, feel that somehow.

When I asked Andrea about the selection criteria, she answered:

> The helper syndrome, the heart in the right place. That is essential. . . . If one's willing to work but maybe not trained professionally, that doesn't automatically mean rejection. Because it's empathy that one must have, willingness and kindness. . . . Why does she want to do this work. Is it about money? Or are they pensioners who really want to be active and help?

Similarly, Melvin, the general manager of the same agency, described the profile of care workers as pensioners who would still like to be active and earn some money in addition to their pension or other income sources. Hence, live-in care work is framed as a side activity that is motivated by a commitment to help someone else. Daniel also compares the care experience of care workers with their motivation and distinguishes between care work in a hospital or a nursing home and live-in care work in someone's home. 'An important criterion is "warm-heartedness" (*Herzlichkeit*) and the will to help someone', he explained. For him, the applicants should understand that being responsible for 20 people in a home is not the same as caring for one single person. He found that lateral entrants were sometimes more suited to live-in care work than experienced professional care workers.

Two points implied by the recruiters in these quotes are to be noted. Firstly, professional health care skills and experience are secondary to the motivation of potential care workers. Secondly, the motivation for live-in care work should derive not from financial interest but rather from some kind of inner urge to help the elderly. The ideal care workers were depicted as motivated and selfless helpers with a heart for the elderly, as opposed to someone who Pascal and Dominik described as 'wage earners' or 'those who smell money'. What becomes clear here is the transnational inequality that underpins live-in care work. The binary classification of applicants into whether they are willing to work or supposedly only after money seems to be connected to the care agents' perceptions that wage standards in Switzerland are significantly higher than in the recruitment countries. This is paradoxical: On the one hand, care agencies rely on the assumption that the monetary incentive of live-in care work is strong enough to compel applicants to work as live-in care workers and to travel frequently hundreds of miles between Switzerland and their own homes. On the other hand, care workers are not supposed to ask about salary at the beginning of a job interview.

This double standard on salary as motivation for care work exacerbates the underlying problems of care and domestic work as undervalued and underpaid employment, which ultimately benefit employers and care agencies (see also McDowell 2009). As Akalin (2015, 72) observes, 'if the term "unskilled" denotes work that "anybody can do", then the question necessarily becomes: why search for the ideal domestic worker with specific qualities?' The definition of care work as not requiring specific skills can be understood as a legitimisation of the low salaries of care workers and, according to Akalin (2015, 72), 'is an intervention on the part of the employers pursued precisely to increase yields from their workers in every sense'. Moreover, the image of a warm-hearted, kind, and selfless person seems to envisage the care worker as a family member rather than an employee. In this context, Schilliger (2014, 195) has noted that the portrayal of domestic workers as family members serves to legitimise the blurring of boundaries between working and free time and to justify longer working hours.

As older experienced women

A second key finding concerns the recruiters' understanding of care work as a gendered activity. Their recruitment practices can be located within existing conceptions of gender norms and the distribution of roles within a household. As shown in Chapter 2, domestic and care work in private households, constructed as women's natural domain, has largely been, and still is, often unpaid and undervalued in contrast to paid work in the public sphere assigned to men. As Lutz (2005) states, this differentiation between what is considered supposedly proper gainful work in the public sphere and care work at home in the private sphere presents an informal gender contract. And the redistribution of care work to another woman is accepted to a large extent, because it remains within the logic of 'doing gender' (Lutz 2005). Correspondingly, most agents would mainly recruit women as care workers. However, the interviewed care agents only vaguely answered the question why women are considered more suitable for live-in care work. Dominik explained: '[they are] just more warm-hearted [*herzlicher*], the mothering instinct, I don't know what kind of thinking, I mean, if you look at old age homes, there too it's women care workers. That's why'.

Even if the agents themselves did not subscribe to an understanding of care work as women's work, they explained that care recipients and their family members would prefer women as care workers. Yvonne, managing director of a placement agency that also offers mobile care in addition to live-in care, states:

> Most elderly women won't accept a man. That's generally a problem in health care. We've employed two certified male nurses [for outpatient care services], but it's significantly more difficult to send them to women. They're two gems. I don't care [that they are men], but I'm trained in the health sector. Try to sell that to an 80- or 85-year-old woman, that she's going to be cleaned by a man.

Hence, although Yvonne personally did not believe that care work should be gendered, her agency still only recruited women as live-in care workers.

What is striking is that many care agents prefer to recruit women of a certain age. 'Most of our women are between 45 and 60. We don't have women under 30. Well, we had bad experiences with some. And if they've got young kids, then they're also less worth thinking about', said Dominik. Apart 'from a human point of view, so that they're not separated from their children for a long time', as explained by care agent Livio, another reason for employers to avoid women with children under 17 years is that they are entitled to family allowances, which employers have to pay. This would make live-in care more expensive. Moreover, Dominik said, 'we noticed that their warm-heartedness has also been lost since the disappearance of socialism. . . . Younger women don't have the same motivation, the same attitudes to family values, which is usually so greatly emphasised in Eastern Europe'. The care agent's explanation here is that 'women of a certain age' would treat the elderly differently than younger women. Moreover, some of the care agents implied that older women would prefer to lead a quiet life at home, while young women would want to go out and have a social life. I was also told that older women usually have a family back home and so would not intend to settle permanently in Switzerland. Care agent Livio even used age to justify difficult working conditions, implying that older women would be able to better handle rough working conditions than younger women because of their life experience:

> Our care workers, they're mature [*gestandene*] women. Don't imagine they're young girls. That may be the fantasy of some public officials. But these are mature 50-year-old women, they've raised children, maybe gone through a second divorce in Poland. . . . They come from very tough private circumstances.

According to Livio, public officials would try to prevent live-in home care and impede the development of young care agencies by prioritising the protection of care workers' working conditions, because such officials have an inaccurate idea of care workers and their capacity for dealing with difficult working conditions. Therefore, Livio argues here that the recruitment of older women legitimises the difficult working conditions of live-in care workers in Switzerland.

Taken together, men and younger women's chances of finding employment as live-in care workers are substantially slimmer than for older women. It must also be remembered that most of the applicants are women. However, Daniel observes that more and more men apply, 'with a background from the health sector, from rehabs, masseurs, or whatever, paramedics, who say, okay, I can earn more salary'. As indicated previously, not all of the care agents agreed with this discrimination against men in the care labour market, especially those who have experience in health professions and the health sector. Yet, as some of the care agents explain, if they do select younger women or men, it is for work in rather atypical care arrangements. Examples include care recipients who do not require around-the-clock care and care recipients who require a physical strength that men would supposedly be better able to provide. Overall, the evidence of the interviews shows that care agencies reproduce the discourse of care work as a gendered domain by predominantly selecting women and, in our case, women of a certain age.

As transient migrants temporarily separated from their own families

The third key point concerns where care workers are recruited. Whether care agencies are located in Switzerland or in the recruitment countries, whether they work with a recruitment partner or recruit directly, they all have one thing in common: None of the live-in care workers they recruited are located in Switzerland but are portrayed as migrant women who do not intend to settle in Switzerland. Live-in care arrangements are clearly not meant for local care workers; as care agent Livio puts it:

> Imagine, a woman comes here as a live-in care worker, starts to save money, and starts to pay rent for her own apartment. And then her boyfriend joins her . . ., they live together. Maybe they have a couple of kids. Now, . . . is this woman supposed to sleep and live in the care recipient's house? Forget it! . . . This model is not suitable for [permanent] residents [in Switzerland]. . . . As soon as they are permanently here and they live here, they're never going to accept a job like that.

Hence, live-in care work is supposedly irreconcilable with having one's own family in proximity. The care workers are portrayed as migrant women who are not supposed to centre their lives in Switzerland, but in what is conceived as 'their own home' in the recruitment countries. Correspondingly, the salary that live-in care workers receive is not usually sufficient to live in Switzerland. As a consequence, and because live-in carers usually have to be on call around the clock, care workers lack the chance to build social relations outside the care arrangement or for social mobility (see Chau, Pelzelmayer, & Schwiter 2018). In this sense, live-in carers are encapsulated by the specific characteristics of 24-hour care work. In periods when they are not on placement, care workers are expected to wait 'at home' in their recruitment countries. 'Because, it's too expensive to stay here [in Switzerland] without income. And why should they, their family is at home in Slovakia anyway, or in Poland', Dora explained. She also differentiated between the recruitment of care workers by outpatient care agencies (*Spitex*), who mainly employ care workers from Switzerland, and the recruitment of live-in care workers. 'I only recruit foreigners', she said. Therefore, the Swiss home care market is generally divided between local care workers for mobile care services who visit care recipients to deliver hourly care and circulating migrant care workers providing live-in care. The recruitment countries mentioned by the care agents and that I noticed on care agencies' webpages were chiefly Slovakia, Poland, Hungary, the Czech Republic, Lithuania, and Eastern Germany. Moreover, care agents reported that they generally received applications from the EU accession countries and from Romania, Bulgaria, and Russia. Hence, the recruitment countries are all located in lower-waged countries east of Switzerland. As my research colleague Pelzelmayer (2016) states, the othering of places where live-in care workers come from serves to justify and sustain inequalities in live-in care work, such as the low salaries of care workers and their difficult working conditions.

Legitimising recruitment from Eastern Europe

In order to further legitimise the recruitment of nonlocals and of women from places in Eastern Europe, some agents even go as far as ascribing specific characteristics to people from these places. Pascal, for example, seems convinced that

> people from the Eastern countries have a different moral concept than we do. Sometimes they are just salary receivers, they wait until the last day, until they can take their 1500 Euros home. The Hungarians, they're the closest to us as far as ethics and values are concerned.

He seeks to recruit care workers with supposedly similar ethics and values to his clients, the care recipients. Hence, he believes that certain similar values between care workers and care recipients are required to create smoothly functioning care arrangements. In this case, Hungarian care workers would supposedly embody a work ethic more similar to Swiss people's than those of people from other places in Eastern Europe.

Dominik compares potential carers from different places by referring to the recruitment of another agency:

> I know that the largest live-in care agency recruits from Eastern Germany. For us Swiss people, High German is a little bit arrogant. And the Germans are a little bit more quick-tempered, faster, and arrogant. They are not as warm-hearted as people in the Slavic parts of Eastern Europe. A typical German also comes across as rather distant and cold.

While care agent Andrea mentions language skills as one of the most important aspects, Dominik deems a 'warm-hearted character', which he associates with the Slavic part of Eastern Europe, more important than language skills.

Finally, the physical appearance of care workers can matter in selection. Livio and Pascal both stress that care workers should have a neat appearance. When Dominik was looking for a French-speaking care worker, he claimed to have faced difficulty in finding a suitable candidate from his usual recruitment countries in Eastern Europe and complained about receiving applications from women of Northern African heritage from France:

> And then, of course, North Africans applied. The 70- and 80-year olds, if someone is black or Asian, then they have different reservations towards them than our generations. Luckily, I still had one, a Polish woman who had immigrated to southern France (. . .) But well, the first care worker was a Slovak, and the old person was very happy with her. And now with the Polish care worker, it doesn't work as well as before with the Slovakian. And her son even said,

maybe, she's stayed too long in France, so maybe she's lost her eastern warm-heartedness, she has more of this arrogant French 'grand nation' attitude now.

Here, Dominik implicitly addressed the skin colour of care workers, implying that he did not take non-white care workers into consideration (see also Liang 2011, 1825–26 for stereotypes regarding the skin tones of care workers). This statement shows that the transnational recruitment of live-in care workers is clearly grounded in complex discriminatory and racist ideologies. Racism here should not be understood simply as an individual attitude and practice; I consider racism as a characteristic of a society that distinguishes people according to characteristics such as origin, recruitment place, and skin tone, and classifies them into groups so that they receive different access to social resources and participation. In this case, these recruitment practices lead to uneven access to live-in care work, because white women from Eastern European countries are constructed as more suitable for live-in care work in Switzerland than non-white women.

These othering practices in the recruitment process also reflect long-extant discourses on Eastern and Western Europe. Melegh (2006) claims that an 'East–West civilizational slope' locates civilisation in the geographic west, so that individuals and institutional actors position themselves as superior to their neighbours in the east but inferior to those in the west (see Melegh 2006 for an extensive analysis of discourses on Eastern and Western Europe). Kuus (2004, 473) argues that the EU enlargement is 'underpinned by a broadly orientalist discourse that assumes essential differences between Europe and Eastern Europe and frames difference from Western Europe as a distance from and a lack of Europeanness'. As a consequence, the EU accession states are depicted as learners of European norms (Kuus 2004, 473). In this sense, if care agencies look for carers from specific places partly because of characteristics they ascribe to women from these places, even if it is merely a legitimisation or marketing strategy, they contribute to a place-related essentialisation of the care workers' characteristics. By actively constructing the association of nationals from Eastern Europe with a character that is warm-hearted and selfless, with a work ethic as close as possible to Swiss work ethics, they reproduce an otherness that marks care workers from Eastern Europe as supposedly more suitable for elderly care work than local care workers in Switzerland. What is especially interesting in Kuus's analysis is that otherness is 'not only inscribed on East–Central Europeans by the West but also appropriated by East–Central Europeans' (see Kuus 2004, 479).

(Co-)construction of care workers as warm-hearted, selfless women

'Recruiters play an important role in the transformative and performative practices of the self that are implicated in the production of live-in care workers', state Findlay et al. (2012). This becomes most evident in one of Daniel's statements:

If the person does not conform [to the requirements], then we communicate that. We show the potential employees very clearly what they can do, so that they can meet the criteria to be employed. Some are grateful; others are disappointed and leave.

Evidently, the recruitment and selection of care workers is a collaborative process in which applicants co-construct their appearance and behaviour in light of their pre-existing understanding of good care work. Many care workers I met stated, for example, that not just anyone would be able to perform this job; in Krisztina's words, 'you have to put your heart into this care work'. Sandra explained that 'only those who feel something should do this [work as a care worker]', and hence, implied that care work is emotional and interactive work.

Women that are looking for a job in care work do not just give up on their first attempt. As we have seen, they try multiple channels and learn how to apply to care agencies. As Rodriguez and Schwenken (2013, 378) indicate, 'migrants make themselves fit into a labour market niche and therefore highlight qualifications and characteristics that are desirable for this niche, while hiding less desirable ones'. Care workers become increasingly aware of the requirements of care work as they become more experienced in finding care work. For example, Krisztina knew that 'those that aren't suitable for this job, that can't integrate, only look for their own interests, who are too aggressive, don't treat the elderly as they are supposed to – they're selected out'. Consequently, those who are not deemed fit for live-in care work are not employed again, especially, as Krisztina explained, if a family reports a negative experience to the intermediary.

Some of the care workers expressed pride in their care work, for example in being able to put the needs of the care recipients before their own and to make the care recipients feel good. Patience, tolerance, understanding, and devotion to work are some of the characteristics that my interview partners mentioned in relation to the requirements of live-in care work. Krisztina talked at length about the importance of 'soft skills', such as the ability to adapt to care recipients' lifestyles and the ability to really care about the care recipients and establish an emotional connection. As we see later, these soft skills come into play when care workers arrive at the care recipients' households. For now, the degree to which potential and actual care workers are aware of the soft skills that, in their perception, are required for care work matters in their recruitment. Krisztina believes that care workers from Hungary should give their best in their performance as care workers, and, in her words, 'be on top, because that image will be passed on'. Finally, I have heard care workers imply that attitudes towards the elderly would be generally different in certain places in Eastern Europe from that in Switzerland and that the elderly would be more respected. Thus, care workers may actively contribute to the construction of the picture of a warm-hearted and nurturing migrant care worker from Eastern Europe as suitable for live-in care work in Switzerland.

However, some of the care workers also express awareness of the fact that the differences between local care workers and care workers from Eastern Europe do not actually lie in different performance, mentalities, or characteristics. As

Krisztina stated, 'it does not depend on whether I am Hungarian or Swiss or anything', but that '[we are] cheaper than the labour force there [in Switzerland]'. Krisztina's statement detaches the live-in care worker from her nationality and indicates that live-in care work is first and foremost based on the economic reality of unequal salary conditions between Switzerland and the recruitment countries.

Ruptures in portraying care workers as warm-hearted, selfless older migrant women from Eastern Europe

Not all my interview partners agreed that recruitment should be based on these stereotypical criteria. Two of the interviewed agents, Yvonne and Dora, stood out among the sample in making a particular effort to recruit in a more differentiated way. With an indignant expression on her face, managing director Yvonne told me how she encountered stereotypes in live-in care work. When a potential care recipient, whom she described as very wealthy, asked her to organise 'a cheap Polish woman' for her and her husband, she reacted like this:

> I said, excuse me, what? 'A cheap Polish woman' [the potential care recipient said]. I couldn't hold my tongue. I said to her, isn't your husband way too old for that now? 'That's not what I meant' [the would-be care recipient said]. But that's exactly what it sounded like!

Yvonne felt offended by the language of her potential client and implied in her answer that the client was asking for a sex worker from Poland. In the end, Yvonne refused to make her an offer, and the would-be care recipient moved to another care agency. Moreover, she emphasised that care workers should not be portrayed as devoted, but that they should be assertive and able to stand up for themselves.

Similar to Yvonne, care agent Dora finds that the most important selection criteria for applicants are strong communication and language skills. She found the first care workers she recruited within her own social network, who then recommended others. Although she recruits from Slovakia, she explained that this was mainly because of her contacts and network in Slovakia and because she had only recently started the placement agency, and not because of inappropriate applicants from other places than Slovakia. She intended later to use Facebook and other social media for recruitment, if necessary. She was convinced that the Swiss live-in care market is still young and will expand to involve recruitment from other European countries such as Portugal and Spain. Therefore, not all care agents legitimised their recruitment channels from Eastern Europe by attributing essential traits to people from Eastern Europe. Both Yvonne and Dora have strong backgrounds in the health sector and extensive experience in care work. Their examples constitute a rupture in the general construction of who is deemed appropriate for live-in care work. It shows how care agents' practices matter in the construction of who is deemed suitable to be a live-in care worker.

Lack of definition of the profession 'live-in care worker'

Although the portrayal of who is selected as a live-in care worker is relatively coherent across all the agencies in which I interviewed, I have also shown some differentiation in recruitment practices between these agencies and how closely their practices relate to differing business strategies. The key criteria in the selection of care workers in each agency depend on the business development and marketing strategies of each care agency. Within the same labour market, the agency that Andrea works for only recruits from Germany and highlights language skills, while Dominik only recruits from Poland and underscores the 'warm-hearted character' of a Polish care worker, whereas Pascal only recruits from Hungary and stresses the supposed similarity between the work ethics of Hungarian care workers and Swiss people. Each care agency foregrounds different key criteria on their webpages and in information brochures (see also Pelzelmayer 2016). This shows how recruitment practices are very closely connected to individual marketing strategies and business practices. It also shows that the role of live-in care worker is presented less as a profession that requires a coherent set of skills and qualifications than as a 'help' to the elderly in need of support. The hierarchies in selection criteria change with the imaginings, marketing strategies, and recruitment access of the recruiters and care agencies. Moreover, it seems that the various constellations of selection criteria are linked to the background and experience of the agents in the health and care sector. Care agents who stress that anybody is able to do care work, such as Pascal and Dominik – neither of whom have experience in the health care sector – look for 'not too dominant' and 'devoted' care workers. In contrast, care agents such as Dora and Yvonne, who both have strong backgrounds in care work, emphasise that care work is demanding and hard work and requires care workers to have strong communication skills and the ability to stand up for themselves. Thus, the various constellations of selection criteria do not follow a uniform logic across the care agencies, and selection mirrors the divergent conceptions of what professional qualifications a care worker should have.

Conclusion: a gendered migration channel from Eastern Europe into Swiss households

The care agencies either recruited their own care workers or worked with recruitment partners to do so. The care agents worked with the Internet and other communication technologies and/or travelled to the recruitment countries themselves. Home care agencies in Switzerland have not developed their businesses from a blank slate; they can draw on existing ideas about migration and social networks and a long history of migration between Western and Eastern Europe (see Elrick 2009, 36). In the small recruitment villages of southern Hungary, going abroad for work is not seen as anything unusual. One such village was of paramount importance as a recruitment site for care agent Pascal when he started his care agency. Pascal's agency provided another opportunity for prospective migrants to work abroad in addition to existing private intermediaries and care agencies.

His agency was not a distant and anonymous care agency. The care workers knew Pascal through his intermittent visits to the village. They also, perhaps most crucially, knew the person who brought Pascal to the village, Krisztina, who had long experience in elderly care abroad. Furthermore, they knew the coordinator, Irina. As a German teacher in the village and as vice president of the local German minority association, Irina held a prestigious position in the community. Both Krisztina and Irina brought tacit knowledge to connecting care workers with Pascal's potential care recipients. This example clearly shows the continuum between the commercial and social dimensions in the migration infrastructure of live-in care workers.

Care agencies in Switzerland without private contacts in recruitment countries find care workers by working with recruitment partners or employing former care workers to recruit more. Collaboration between care agencies and recruitment agencies is facilitated by communication technology such as Skype, e-mail, and automated notifications when prospective care recipients and care workers fill out forms on webpages. Online communication allows care agents to communicate directly with partners in recruitment countries and to communicate with employees and collaborators outside Switzerland. Such collaboration does not always originate from new home care agencies in Switzerland; it may also be initiated by care workers, former care workers, and recruitment agencies in the recruitment countries. Many care agents in Switzerland told me that they are frequently contacted by such agencies. In such cases, home care agencies facilitate the first step in migrant live-in care work by providing new channels for prospective care workers to find employment. They create new platforms for live-in care workers to access employment via their webpages, collaborate with recruitment agencies in the recruitment countries, and tap into care workers' social networks and communities; they thus provide an essential element of live-in care workers' migration infrastructure.

However, they also function as gatekeepers in care workers' access to the Swiss live-in care labour market. The recruiters select care workers by gender, recruitment locations, age, family situations, and more intangible criteria such as the motivation and mentality of applicants. These recruitment and selection practices characterise live-in care workers in Switzerland as older, warm-hearted, selfless, and caring women from Eastern Germany or Eastern Europe who are strongly motivated to help the elderly. These findings support the construction of an image of such care workers that previous researchers have analysed as gendered, racialised, and ethnicised (Loveband 2004; Guevarra 2010; Liang 2011; Schilliger 2014; Schwiter, Berndt, & Schilling 2014). These findings do not come as a surprise. Many studies of migrant domestic workers have demonstrated the tendencies of employers and placement agencies to racialise and hierarchise domestic workers in various geographical contexts as a means of legitimising a range of practices (Bakan & Stasiulis 1995; Constable 1997b; Huang & Yeoh 1998; Stiell & England 1999; Parrenas 2000, 2001; Anderson 2000; Lan 2006). Bakan and Stasiulis (1995, 317) note that 'racial stereotyping is endemic to the matching process' of placement agencies and that rendering a perfect match in practice refers to an agency's 'ability to stereotype in a way that is consistent with employers' expectations'. In this sense, stereotypes, whether racial or in relation to gender, recruitment countries, and age, serve to legitimise placement agencies' existence; the screening of workers is an essential part of their services (see also Lan 2006). The stereotypes, as Paul (2013, 1073) points out, are 'relational, hierarchical and contextual'. Depending on context, care and domestic workers are constructed as docile, disciplined, and submissive in Southeast Asia (Liang 2011), or as obedient and loyal in Taiwan (Loveband 2004), and in our case as patient, warm-hearted, and selfless.

Hence, by engaging in spatially selective recruitment of workers from certain places and by linking the places, such as Eastern Europe, to certain traits supposedly required for care work, such as warm-heartedness and selflessness, care agencies contribute to the creation of stereotypes and discriminate against would-be care workers from other places. Consequently, their recruitment practices actively contribute to both the construction of a gendered migration channel from Eastern European countries to the Swiss live-in care labour market and the production of an unequal division of care labour in the transnational context.

7 Matching with and travelling to the workplace

'When we do a good job, so when we compare the needs analysis of the client with a personality analysis and find the right match, a care worker, when we can place a person (long-term) and only have to find holiday replacement for the person, then we are happy and the family is happy', said care agent Daniel. What is striking in this quotation is that Daniel mentions the care recipients' needs and the care agents' and care recipients' happiness. The care workers' needs and feelings, however, remain unaddressed. This chapter looks at the continuation of care workers' journeys into live-in care after being recruited by a care agency. First, I present the matching practices of care agents. I provide an insight into the process that defines in which households care workers are placed. Subsequently, I show care workers' experiences of how they travel to their workplaces. I describe how care agents organise care workers' journeys and, in so doing, how they set up an infrastructure tailored to care workers' specific mobility patterns. In the conclusion, I indicate the asymmetric power relations in both the matching process and the organisation of care workers' journeys.

Playing Tetris: the matching process

Home care agencies are a viable option when seeking employment in live-in care work, especially for those who lack social networks with access to employers. However, having passed the stage of application and recruitment and being registered in the pool of a care agency does not yet mean that a care worker actually finds employment. As shown in the previous chapter, few care agencies directly recruit for a specific position. An actual position usually only materialises during the matching process if a care agency finds a household where a would-be care worker can work. 'Matching' is the process of choosing in which households care workers will be placed and thus for which care recipients the care workers will be caring. According to participants in a study by Krawietz (2014), matching consists of bringing together the right people, or to choose people for each other and to find out which care worker matches with which care recipient.

As care agencies recruit care workers, they simultaneously also actively try to acquire care recipients as clients. Once a prospective care recipient makes a request, a care agency starts to organise the arrangement by finding a corresponding care worker who is available for the period requested and interested in working in that household. 'What is requested? What is needed? Which skills do the workers have to bring?' are the crucial questions in the matching process, as recruiter Andrea explained. 'And then we scan our pool. Not everyone fits. . . . It is a little bit like playing Tetris. You have to piece the blocks together'. How do care agencies know which 'blocks' actually fit? The following sections provide insight into how care agents construct their roles as indispensable intermediaries through the matching process.

Standardisation and matching software

Krawietz (2014) shows that care agencies operating in Germany and in Poland use both standardised procedures as well as subject-oriented, personal-interactive procedures to make a match. Similar to Krawietz (2014), I find that care agencies have, to varying degrees, introduced standardisation procedures with questionnaires and CVs for both prospective care recipients and care workers to facilitate the matching. The questionnaires are filled out online or in paper form, usually by care recipients' family members, who, according to care agent Dominik, are often women. Sometimes, the agencies also ask for the information directly over the phone with the prospective care recipients to make sure that they do not back out if they are left alone with the questionnaires. The questionnaires assess the care recipients' health status and care situations, such as whether they have particular medical conditions, special diet requirements, and mobility issues. Moreover, they can address care recipients' preferences for care workers' experience, gender, ability to drive, and permission to smoke. In addition, they ascertain conditions in the household, such as whether there is an Internet connection, how many people live in the household, and whether there is special care equipment such as a hospital bed. In parallel, care agencies also collect care workers' data on experience, skills, age, and so forth and standardise this in CVs. These records are then scanned, and the care workers who are considered suitable are selected. Agencies working with recruitment partners in a country of recruitment usually receive some CVs from their partners on request. The agency then either chooses one or forwards a choice of three to five CVs to the care recipients. The matching procedure here is divided between the care agency, the recruitment firm, and the care recipients and their family members and thus occurs across different countries. In this case, it may be that care recipients are only in contact with customer care agents, who are in contact with recruiters with access to care workers.

In order to manage the lists of care workers and care recipients and the overall matching process, some care agencies use data management software. Whereas some of the agencies interviewed used off-the-shelf products, one of them had asked an IT specialist to create an interface that would facilitate data management. The agency that Andrea works for recruits centrally. Hence, Andrea, who was in charge of the recruitment at the main office, usually received requests from the agency's branches all over Switzerland. She typed the care recipients' data into a program called *Boss*, which filters the care workers. Additionally, she worked with a board on which she pinned pictures of the care workers and kept an overview of who was available for placement and with which branches they were already placed. Since Andrea personally recruited the care workers and saw them face-to-face, she knew each of them to some extent.

At the time of our interview, Andrea explained that she was still trying to figure out how to describe the criteria and skills of individual care workers in written form so that someone other than herself would also be able to work with it. She struggled with this effort, especially in the beginning, as it would be difficult to assess someone 'who happens to appear in a certain way to someone, but appears in a whole different way to someone else'. Over the course of a year, she had been trying to improve the questionnaires so that fewer key words would be needed, so as to improve the efficiency of the matching process. According to Andrea, keywords such as 'especially friendly', 'suitable for sophisticated household', 'smoker/non-smoker' proved to be particularly useful. Before she started using the matching software, she had handled data management of personnel files manually:

> When I started working here, my predecessor managed everything out of her head. She only had thirty care workers and she knew them all. I started to insert data into the system and I'm still working on that. Because I know them all personally, and the others don't. But the purpose is that other employees one day are able to take my place so that they can work with this [system].

Therefore, the software serves not just to facilitate data management and to ensure an efficient matching process, but also to a certain extent to detach or disembody the ability to match care recipients with care workers from a specific person. In other words, Andrea's goal was to transfer her tacit knowledge, which is difficult to represent in symbolic forms and hence difficult to transfer to other people, into codified knowledge, which can be articulated and accessed, and hence easily transferred to others (Gertler 2003).

Before care agent Anthony tested eBoss, an online version of the Boss software, he had worked with Excel sheets. However, with 200 care workers, he felt that he had reached the limit of its function. He appreciated the new software, as it can perform automated tasks such as cancellations, create standardised templates for e-mails, and calculate and record working hours, holidays, and much more:

> You can connect two care workers to one customer. We weren't able to have that link with Excel. So, we do hope it'll be a relief. It's better that we change to something more professional now and not wait until we have 500 [care workers]. . . . You can scan CVs too. I scanned my CV, and the system was able to retrieve all the information automatically.

At a cost of CHF 100 per month, the software enables more efficient work so that, in Anthony's words, 'there is more time for other things'. The examples given here show how care agencies have started to introduce standardisation in their back offices to facilitate working flows. However, the standardisation is performed within an agency and not across agencies. Hence, the degrees and ways of standardisation of assessment and matching vary between agencies.

Matching as interactive work

Although agencies can facilitate human work with the help of computers and software, the matching process can only be streamlined to a certain extent. It cannot be fully automated. In the end, whether a care worker is supposedly 'especially friendly' or 'suitable for a sophisticated household' still has to be assessed by a recruiter before it is inserted into the software. Just as with recruitment, care agents assess a potential match based on their own relation and interaction with the care workers and the care recipients. Hence, matching is always also interactive work. Smaller agencies with a low degree of standardisation rely especially strongly on personal and interactive procedures that focus more on communication. Care agent Pascal, for example, would call Irina, the German teacher he collaborates with in Hungary, and discuss the matching process with her: 'What do you think, which one should we take?' Through Skype, she would call everyone on the list to ask if they were available. Before Pascal contacted Irina, he personally visited the households where the care workers were to be placed. He found face-to-face contact with both care workers and care recipients very important for building his business. According to him, it is what distinguished him from other care agencies and, he believed, was his key to success. Dora, owner of a one-person business, also thought it important to personally visit the household where the care workers live before signing a contract. She attached importance to verifying the conditions required at the house, such as whether the care workers are provided with their own rooms. She also showed the care workers' application dossiers personally to the care recipients before they made their decision. However, not all care agents interviewed emphasised face-to-face contact and interaction as much as Pascal and Dora. Some agencies are even in contact with care recipients and their family members and care workers only online or on the phone. Therefore, care agents invest different degrees of efforts into interactive relationships.

Encouraging physical presentations to facilitate care recipients' choices

One way to make up for sparse or 'thin' interaction between care agents and care recipients is to create a feeling of interaction between the prospective care recipients and the prospective care workers before a placement. Once a care agent, sometimes with the help of matching software, has identified possible care workers for a care recipient, the care agency either determines the final candidate or forwards a selection to the care recipients and their family members. Pascal did not see any sense in giving care recipients a selection but preferred to decide himself who was going to be employed; however, other care agencies I interviewed usually sent the care workers' CVs to the care recipients. One of the placement agencies even sent a selection of videos of the care workers. The care workers were instructed to portray themselves in a short clip. The two videos, which care agent Michael sent me as an example, were between three and five minutes long. In the first one, a 33-year-old woman with glasses and a ponytail and wearing headphones with a microphone briefly introduces her name and age before she lists her experiences and skills:

> I have learned to cook a couple of Swiss dishes. . . . And you can teach me more dishes. I can cook Rösti [traditional potato dish] and Hörnli [Swiss term for macaroni]. I am experienced with different conditions, such as dementia, Alzheimer, strokes, and fractures. I have experience with incontinence, changing diapers, with movements with walkers and wheel chairs, bodily hygiene, shaving, laundry, cleaning, grocery shopping, cooking, playing games.

The video only displays the head of the woman, as she is sitting in front of the video cam recorder or computer in a room. The background is grey and rather dark.

The second video is brighter and was recorded in an outdoor setting. It shows an older blonde woman, 56 years old as she states, wearing a bright pink jacket. While she speaks, I can hear cars driving by in the background. Her presentation seems more prepared and practised. First stating that she had been working as a care worker for the past seven years in Germany, she then talks about her experiences with three care recipients, their different health conditions, and her interactions with them. Subsequently, she talks about her own family and why she would like to work in Switzerland, namely that her son, who used to work in Switzerland, had told her good things about Switzerland and also that the salary in Switzerland would be better for her. She also states a few requirements or hopes for her future working arrangements, before she finishes her video by stating her own hobbies:

> I want to work in 3- or 4-week arrangements, as I live by myself in a 3-bedroom apartment and regularly have to attend to matters in relation to my apartment. My client has to be fit in her mind. The family cannot have pets, because I am allergic to cats and dogs. My hobbies are going for walks and visiting the theatre and concerts.

In contrast to a CV, care workers can address the care recipients more directly in a video, by using phrases such as 'you can teach me more dishes'. In some cases, for example with Dora's agency, care recipients and their family members could contact the proposed care workers and talk to them. If the final selection consists of care workers with similar experience, the crucial reasons leading to the final choice can be difficult to grasp. According to Dominik, the physical presentation of a care worker can play a role: 'In the end it is the picture, or the zodiac or something. That's why we tell the care workers to take a good picture'.

In this sense, care workers are encouraged to appear 'presentable' and to create interactive relations even before starting an employment to increase their chances of being chosen.

Unequal power relations in the matching process

What becomes apparent is the asymmetry in the triangular relationship between care recipients, care agencies, and care workers. The people making most of the decisions in the matching process are the care agents and the care recipients. The care workers have relatively little control in comparison, as after being selected they are simply left with the choice of accepting or rejecting the employment that is offered to them. Unless they apply to a specific post in a specific household, where basic information about location, age, and medical conditions are already known, care workers are usually not presented with a choice of households; they are only presented with one possibility.

Moreover, the employment offer can be made at any time. The care agents usually do not inform care workers when they can expect a position. In cases where care workers are asked to begin a placement as soon as possible, for example if a care recipient is released from a hospital and needs care immediately, those care workers that are flexible enough to accept the conditions and start work the next day have an advantage over those that need time to organise their journey. As we have seen in previous chapters, care agents try to increase a flexible disposition by creating pools and, hence, competition between care workers. Speed matters in the matching process. The higher the readiness of care workers, the faster they can be placed, the more care recipients and care agents benefit. Consequently, those that are more flexible can actively weaken the possibilities of migration for care workers that are less flexible. As Chapter 5 shows, these less-flexible workers usually have less experience and families at home. Furthermore, it is not just other care workers that can be affected by this demand for readiness but also the families of the care workers themselves, who have to support this flexibility and organise their everyday lives accordingly.

Mystification of the match

It is important to note, however, that although the power to match families and care workers lies with the care agents and the care recipients, the final say on whether the match is made lies with the care workers. Care agents have preferences in the selection of care workers and hope that the care workers they select agree to the match. Pascal explained that 'most of the time, those we do not want are available and those we want are not. But then she [Irina, the German teacher] tries to convince them within one or two days, come on, go now, we just need you now'. To Pascal and many of the interviewed care agents, it matters which care worker is placed in a specific household. Although the care agencies' business models are based on the disposability of care workers, it does not mean that they are interchangeable in any situation. If that were the case, then care agents would lose their legitimacy as 'indispensable' brokers, a role they try to create not only with the provision and building of professional recruitment through standardised screening and matching infrastructure with software programs, questionnaires, and so forth but also by emphasising and mystifying their own recruitment and matching skills. The interviewed care agents emphasised how crucial was their ability to assess the potential between the chosen care workers and the care recipients to build a good relationship. 'Matching is psychology', Dominik stated. Some of the care agents explained that their own private and professional experience was relevant to making the right match. Yvonne had personal experience with taking care of her own mother and had been working in the health sector for years. She argued that 'you cannot learn life experience, you either have it or you don't'. Dominik underlined his experience and knowledge from his other business, a temporary staffing agency, and referred to his experience as a debt collector before starting his placement company: 'I seized things from people for seven years. I took away intimate things from them. So, that was a very social job, so I have a sensitivity of who could fit'. Although I would not call debt collection a prosocial occupation, I understand what Dominik wanted to say: that it would require social skills to execute debt enforcement. Moreover, care agents stress that 'talent', a 'gut feeling', and 'seven senses' help decide matches. Pascal even claimed that the matching process cannot be explained or standardised. Whether a care arrangement works out would very much depend on his recruitment and matching skills. Hence, by describing a match as 'inexplicable' and mystifying it, care agents legitimise their role as intermediaries.

However, care agents are also aware of the uncertain component of a match. Whether a match works out also involves luck. However carefully care agents prepare a match and choose a care worker, there is always some uncertainty. As Dora puts it:

> You can't know. Even after talking to somebody for two hours, you can't tell for sure that they are the one. You can influence a lot with experience and conversations, but let's just say about thirty of forty per cent is just life and how that develops.

According to a study by Van Holten, Jähnke, and Bischofberger (2013) on the family members of care recipients in live-in care arrangements, this view is shared by care recipients and their family members. The authors found that the family members they interviewed perceive a well-functioning care arrangement not as a natural process but rather as a lucky exception (Van Holten, Jähnke, & Bischofberger 2013, 30).

Travelling to work: care workers' perspective

'The trip was a little bit complicated' is how Krisztina described her journey to her workplace in Germany, where she used to work as a live-in care worker before she found her current employment in Switzerland. To reach the household where she worked, she would leave her village at 7.30 in the morning and arrive more than 20 hours later. She would take a train from her village to Budapest, where she would take a bus to Prague, change to another bus to Hannover, and finally reach the town of Celle the next morning. Labour migration is often associated with long journeys and more permanent stays rather than with commuting. In Krisztina's case, however, there is a certain frequency in her travels. The movement of live-in care workers going back and forth between home and workplace in regular intervals has been conceptualised as *Pendelmigration* by German speaking researchers, the media, and state institutions (Ellner 2011; Fürst 2013; Medici & Schilliger 2012; Schilliger 2013, 2014; SECO 2012; Strüver 2011, 2013; Truong, Schwiter & Berndt 2012). Translated into English, the term literally means 'commuting migration'. It has been argued that live-in care workers commute in this way not to migrate to the country of the workplace but actually to enable their periodic return home to their recruitment countries (Schilliger 2014, 290; see also Morokvasic 1994).

This temporary and repetitive mobility pattern of live-in care workers in Europe is also known as circular migration (Chau, Pelzelmayer, & Schwiter 2018; Marchetti 2013; Triandafyllidou 2010; Triandafyllidou & Marchetti 2013). The way that care workers travel between residential location and work place is a crucial part of the underlying inequality that is inherent to live-in care labour migration. Whereas research on a more traditional understanding of commuting has shown that full-time working men with higher incomes travel more and longer distances than part-time working women with lower incomes (Crane 2007; Hanson & Johnston 1985; Madden 1981), our study provides contrary findings. Care workers regularly undertake long and sometimes cumbersome travels to reach their workplaces. How is this mobility achieved? How do care workers travel to their workplace and back home? This section provides insight into the various ways in which care workers travel from their own homes to their workplaces and how care agencies organise the transport of care workers. The following examples refer to journeys to places in both Switzerland and Germany.

Different starting points in organising the journey

Travel to the private households in which care workers start their employment in live-in care in other countries is organised in one of three ways: 1) by care workers themselves; 2) by home care agencies; or 3) by informal drivers and brokers. If care workers have to cover the costs by themselves, the costs can play a role in their choice of transport. 'Some families pay for the journey, some don't, so in this case, you just look for the cheapest option: the cheapest option between train, minibus, or plane. You can save money this way', explained one of the care workers that I met during fieldwork. When I visited Anna in Harta, she took my fieldwork collaborator Katalin and me to a café where a group of care workers met in the afternoon. The care workers estimate whether their transport is worth it or not by comparing the different transport systems, as the following extract from the group's conversation shows:

NORA: I get 1,350. And the first group that arranged the travel asked for 100, 110 Euros. Now I found someone who takes me to the same place for 85 Euros.

LENA: The Szekszárd group is very cheap. If I travel back with them, they ask for 300 Euros.

KATA: I tried Eurolives, I travelled with them in the first three years.

ANNA: I fly by plane.

NORA: How many kilometres are you away?

ANNA: (answers, but we can't hear the answer).

KATA: I travel to Frankfurt, it is 1,000 km.

NORA: How many? Oh my god 1,000? That's really far! (yells)

ANNA: I fly through Frankfurt to get to Zurich.

KATA: It's cheaper if you go through Frankfurt, but it means that there's another 100 km between Frankfurt and the airport itself. Actually, I pay 60 Euros to Frankfurt, which is a joke. That's so cheap. With this small van. Their name is Europaclipper, which is a Slovakian business. Slovak-Miskolc business [*Miskolc is a Hungarian town*]. Because starting the business in Slovakia is a lot better. They have less tax, which is why a lot of businesses are registered in Slovakia. They pick me up in Keleti Palyaudvar in Budapest at 9 a.m., and we get to Frankfurt around 8 p.m. They go directly. Sixty Euros, 1,000 km, that's nothing.

NORA: I pay 85 Euros, but they pick me up directly in Kicsifalu.

KATALIN: Do you all travel by bus? Are you all in favour of the bus?

EVERYBODY: Yes, yes, by bus (*except for Anna*)

NORA: Me and you (*not sure who*), we're transported by a man from Kicsafalu. He got the job for us too. They even lift my baggage. I don't even have to lift my own baggage. It is directly from house to house. In Kicsafalu it is like a kind of little business. There's a couple that gets the job for you. And their relative does the transport. So, it is a little family business, but there is nothing wrong with that. The most important thing for us is to have a job.

Not only does the extract show that care workers share knowledge on different means of transport in their everyday life, but it also demonstrates their different starting points to organising their journey. For example, Anna took a flight from Budapest to Zurich to reach her workplace. At the beginning of her employment, she took the night train several times. In contrast, Kata takes an 11-hour bus ride to reach her workplace in Frankfurt, which is as far away from Budapest as Zurich is. With a daily rate of CHF 120 salary and an additional CHF 600 per month that her employer in Zurich agreed to contribute for travel costs, Anna was willing to spend more for more comfortable transport than Kata, whose salary was much lower at 40 to 50 Euros per day. Therefore, the choice of transport depends on whether the family pays the travelling costs in addition to or as part of the all-inclusive service or whether the care workers pay for it themselves.

How social relations are embedded in economic practices

The women continued the discussion by talking at length about the numbers of kilometres between different places, such as Budapest and towns in Germany and the different transport options. The travel can cost more if it is organised privately by an agency or by a private broker, such as a transport business. The latter commonly charge a fee for their mediation service and require care workers to travel with them. The care workers' flexibility is then restricted: 'If I say I don't want to come home now or only for two days, I can solve it in a private way. But with them [*transport company*] I can't do that', one of the care workers said. Alana, another care worker in the conversation, recalled:

> One time I had to replace someone and I had to travel with that driver. The driver said 100 Euros to the family. But it was in Munich! It's absurd. That's too much. So, the driver got 80 Euros and he gave me 20 Euros. You can generally get to Munich for 40 or 50 Euros.

In this deal, the driver charged the family a higher price for transport than care workers would pay if they travelled independently, and he gave the care worker a share of his profit. This kind of practice seems to be familiar to the care workers, as Alana explained:

> That's how he starts: yes, I can take you, but it means that you have to travel with me. You have to say this and this amount to the family and then you will get 20 Euros. Back and forth, that's already 40 Euros.

Hence, in exchange for matching a care worker to an employer, in this case the private broker not only binds the care workers to his transport service but also asks the care worker to collaborate in overcharging the employer. In return, the care worker receives a fixed commission.

The landscape of transport businesses and intermediaries and the travelling costs are not fully transparent, either to me or to the care workers I talked to, as the following extracts from the conversation show:

ALANA: I've heard about them, but I don't know them.
KATALIN: Yes, we wanted to contact them, so if you have any information?
NORA: Whom? Internet?
KATALIN: No Interland. Those who provide transport from Hajos.
NORA: Aaah, the Hajos agency? We don't know.
GABI: My mum used to travel with them, a long time ago.
ALANA: It is not Horváth's, but another group. I forgot their name. There are so many transporters. This guy, I don't know the name, can get you places, (later)
MARIE: We pay 80 one way? Those who are from further away pay the same amount. My switch partner is from a place further away, but she pays the same amount. With Tibor. He is a driver.

NORA: Oh, are they with Interland?

ALANA: He is a driver from Baja. I think they are with Interland.

MIXED: Might be, I don't know, well (no consensus, whether that person is from Interland or not).

KATALIN: Do you perhaps have contact with him?

CHAOS. Some people say no. Someone said: He's a friend of mine online.

(GABI: Anna look, this is my oldest grandchild.)

ALANA: I'm sure he won't volunteer for such an interview.

KATALIN: Do you think, we can call him?

ALANA: No, I don't think so. For sure not. It's my personal opinion, but I'm sure he wouldn't agree.

KATA: Yes, because they do it 'black' (*work informally*).

KATALIN: Actually, we don't care if they work informally or not.

KATA: Yes, I know, but not everybody is a smart person to understand this. (a couple of minutes later)

KATALIN: (To Gabi) You mentioned your mother used to work with Interland. Do you know anything about them?

GABI: I said that my mum travelled with people from Hajos. But we don't know with whom. And now we think they might be the same.

As the extract shows, there are many options to organise the journey with private transport businesses and private individuals that mediate care workers informally. The transport company Interland, with which Krisztina found her employment in Switzerland (see Chapter 5), also came up in the group conversation in Kicsifalu, initially because Katalin and I asked about it. Kicsifalu is about 100 kilometres away from the village where Krisztina lives. Before she changed to Pascal's agency, Krisztina travelled by bus with Interland for one year to Switzerland. For her journey back to Hungary, she had already changed to Pascal's transport organisation. However, Krisztina still had to pay Interland for both journeys. The fee was 300 Euros. She explained:

> I'm aware of this audacity. Still, I didn't want a conflict with them, because you can never know when I'll need them. I told him: Alright, I don't think that this is fair, but I will pay, because who knows if I'll need you sometime, so I don't want to part in anger. I will pay the 150 Euros out of my pocket.

Her example shows how the transport infrastructure is deeply embedded in social relations and efforts to stabilise them. Although she felt that it was unjust of Interland to charge her money, she chose to maintain a good relationship with the business, which in the future could be of importance to her.

Weighing convenience, comfort, and costs in the choice of different transport means

The care workers I met in the café in Kicsifalu are very much aware of the advantages and the disadvantages of traveling with a home care agency and/or private transport businesses. 'It is a great help for those who use the service of an agency. But this help has its costs. You are very constrained'. If a care worker has to travel with a care agency or with informal brokers, they may also incur other costs or inconveniences such as longer traveling times and reduced comfort: 'And they go around the towns. They stop in between. They don't go a direct way', Alana said in the group conversation. 'Imagine, they picked me up at 4 p.m. . . . [and then in the] morning at 4 a.m., I arrived. They took me all over the [place]', she recalled of her experience with an informal driver. Kata explained why these cars take such a long time to arrive at their destination:

> So these people usually work in a way that during the trip they get a call. And if there is still free space in the bus, and it does not matter if you have to travel extra, a hundred kilometres, they would still add people on the way, because that means extra money. ⸳

She prefers to travel with a transport business called Europa clipper, which is officially registered and which only provides transport. Having travelled 26 times with that company, Kata appreciates that they would start in Budapest and drive directly to Frankfurt without taking a detour of a couple of hundred kilometres.

In addition to the transport of the care workers, the private drivers sometimes also use their tours to exchange goods and transport them from one place to another: 'When [I] go home, I am usually the first one who will be picked up on the way back', Alana continued her explanations and talked about who fits in the bus: 'More and more people then get on the bus. There are a lot of things, like chairs, other objects that are carried in these buses. That's why the luggage sometimes won't fit'. To address the problems of luggage space, it seems that transport businesses often impose a limit on the luggage. One of the care workers in the group conversation explains that some transport businesses charge them 5 Euros if they bring more baggage than allowed. The following Facebook post by a care agency urged the care workers to bring less luggage (Figure 7.1):

> Our dear care assistants. First of all, I would like to thank you for your great work. However, we have a big plea, or better say request, for you. Here is the photo of a car traveling from Switzerland. As you can see the car is fully loaded, because some of the care assistants even had three suitcases. And do you know what happened? There wasn't any more space in the car for the last carer. So we had to organise another transport for her. Fortunately, we made it, but the carer was picked up late at 7 p.m. I don't think that you would like it. Because of that, we would like to ask you to bring max 1 big suitcase plus smaller luggage, so that this situation never happens again. Thank you!

Figure 7.1 Facebook post by a care agency, October 2014.

Source: (Translated from German into English by author)

Care workers reacted to the post with these comments (names changed):

AGNESA: If they are traveling from Switzerland, I am not surprised that they have so many suitcases. They don't travel home every two weeks. The taxi [transport] company should think about it, not the care assistants.

GABRIELA: The transport company should ask about the suitcase first and after that they can arrange it.

IVANKA: Maybe I saw this car close to Bratislava and I was asking myself if it's even safe. On the other hand, three months is quite a long time.

KLARA: I travel from Germany and I go for a long time. The travel company I was traveling with had the same problem. They solved it that when you called them, they ask you how many suitcases you have. The first piece of luggage is for free; any other pieces are 9 Euro. Then they think about what kind of crap they take with them. For example, one of them took stones, because they would be nice in her garden.

HANA: I travel from Austria and one time a driver asked for 4 Euro because I wanted to take one box of softener. Even though I had the smallest luggage (could not have been even 7–8 kg), and another girl had such a big suitcase, that I would be able to fit in with my 80 kg. If they want it to be fair, they should . . . weigh the suitcases.

The comments show how the organisation of the travel from home to workplace involves everyday practicalities such as decisions on what and how much to bring and negotiations about luggage space between care workers, transport businesses, and care agencies. Furthermore, the comments demonstrate that a journey to Switzerland is associated with a longer stint of three months. Switzerland is further away than, for example, Austria, and hence, it seems, the commentators sympathise with care workers who might pack more luggage for a longer stint.

To summarise, there are a multitude of ways to travel back and forth, including public transport, low-cost private drivers, and more professional transport services that travel from city to city. The care workers try out various means of transport and gather and exchange their experiences with costs, travel comfort, and the way the different transport businesses work, not just by word of mouth but also online. Their choice of travel depends on numerous factors, such as their financial situations, contacts with private drivers and transport businesses and obligations to travel with them, and whether care agents and employers pay for and organise their travels. I also remember vividly how Mira, a woman from Poland who used to work as a live-in care worker in Switzerland, enthusiastically told me in an informal conversation that the low-cost airline EasyJet was now flying from Switzerland to Warsaw. The connection made it much easier for her to visit her home. However, where public and private transport infrastructure is not sufficient or affordable, care workers use informal and private transport businesses that cater to their specific needs.

Organising the travels for care workers: care agents' perspective

On our way back to Budapest from the little village in southern Hungary, Katalin and I had to change buses in the neighbouring village. Since we had a little time in between, we decided to take a stroll. It felt nice and relaxing to walk along the streets after the intensive interviews that we had conducted in the hours before leaving. It was a sunny afternoon in late July and pleasantly warm. There were many fruit trees along the road with overripe plums and pears. The atmosphere was relaxed. Just before we arrived back at the bus station, a white mini bus with two men in the front rode past us. It had a Swiss number plate from Zug. 'That must be their bus!' I yelled at Katalin, referring to the bus that was about to bring the care workers we had talked to the same morning and the day before to Switzerland. Although I made that connection very quickly, I remember how my brain still had trouble connecting what I had just seen and how puzzled I felt. We did not expect to encounter a minibus registered in Switzerland as far away as southern Hungary at that moment. In the next sections, I explain why that bus with a Swiss number plate passed us in a tiny village in southern Hungary and the various ways in which care agents organise the journeys from recruitment countries to their workplace in Switzerland.

Setting up one's own transport system

The bus that had surprised Katalin and me on our way from the little village in southern Hungary to Budapest belonged to Pascal. His care agency was one of two care agencies interviewed that set up their own transport system for care workers registered with their agency. He explained:

> I collect them along the line from southern Hungary, right next to the Serbian border all the way up to Budapest. That line. I don't care who calls then, if I can somehow pick them up on that route. Some have to travel themselves for a little bit, but we collect them on that route.

In individual urgent cases, that is, if the care recipient's family explicitly requests an earlier placement, Pascal also arranges a flight for the care workers. However, in normal and non-urgent cases he arranges the start of a placement and the change of care workers according to his regular bus schedule, or in his words, on 'official switch' days. Every fourth Friday evening, the bus leaves that little town and arrives in Switzerland the next day. In Switzerland, the bus brings the care workers directly to their households. At around 7 or 9 p.m., it is back at the border and on the way to Hungary. Since the drive is long, Pascal feels that it would not be safe with just one driver, so he has hired two. One of the drivers was Krisztina's son. 'He works for the police. So, we don't have any problems, if a car with Swiss number plate is in Hungary for the whole month. He takes care of that. If not, one could get into trouble'. It was not very clear to me what Pascal meant by 'getting into trouble'. However, clearly Krisztina's son's place in the local community somehow played a role in Pascal's smooth transport of the care workers. To make sure that the driver did not drive extra miles while carrying out these tours, Pascal instructed him to send a picture of the tachometer every Sunday evening or Monday morning after arrival. He felt that he had to control the drivers' activities; if not, he feared that they would start to offer private rides. 'And that would not be funny', he concluded.

Organising travels with public transport and existing transport infrastructures

Whereas most interviewed care agencies work with transport partners and only organise flights in cases of emergency, one of the care agencies differs in that it uses public transport. Most of the care workers of this placement agency are recruited from northern Germany, and the agency books flights for them. Although care recipients pay up to CHF 12,000 per month for live-in care with this agency, it deducts CHF 200 from care workers' salaries per stint for the organisation of the journeys. After their arrival at the airport in Switzerland, they take a taxi to reach the household. Hence, care workers that are employed with this placement agency have shorter travel times and travel more comfortably than with most of the other agencies. However, they also do not have the option of using less expensive journeys even if they wanted to.

Establishing collaborations with transport partners

The other agencies interviewed work with transport partners to organise the care workers' journeys. One of the care agencies works with transport businesses in

Poland that brings care workers to Switzerland by bus. However, 'depending on our customers' wishes, we also organise flights of course. That differs from client to client', the care agent Daniel said. It was not fully clear to me who can finally decide on the means of transport and in what way this decision is negotiated between care recipients, the care agency, and care workers. However, it was striking to me how Daniel mentioned the care recipients' wishes but did not mention the care workers' wishes at all. After all, it is the care workers who undertake the journey, not the care recipients. This finding points to unequal power relations, in that the care recipients can play a crucial role in the comfort and costs of the care worker's journeys.

Care agent Dominik came to know that some transport businesses drive from Slovakia to Switzerland from his Slovakian wife. The first transport partner he worked with was based in Germany. It was a professional transport business with around 20 car trips per month all around Europe. The partner picked up the care workers in a nine-seater car at their homes and brought them to the doorstep of the households where they began work. Depending on each individual care arrangement, the drivers waited 10 to 20 minutes until the care workers arranged the changes or they drove on directly after drop-off and picked up the returning care workers later on their way back. As this transport partner only drove on fixed days, for example, only arrived in Switzerland on Tuesdays, Fridays, and Sundays, Dominik looked for another partner. He also wanted to provide more comfortable rides for the care workers: 'Maximum four to five care workers in one car so that the car is not full. Because, they sit for 20 hours like sardines next to each other [with the German transport business]'.

In his search for another company, the first difficulty was in being able to find a partner that could speak German. He came across a Swiss man who lived in Slovakia and started to work with him. However, the collaboration did not work out, as Dominik explained:

> somehow that guy did not pay the drivers properly. And then suddenly the owner of the buses called me and asked me if I was happy with that guy, because, apparently, he had not received any money yet from him [the man Dominik collaborated with], and, apparently, he didn't clean the bus. It turns out that guy didn't have any busses of his own.

Hence, Dominik kept looking for a new partner and found another small business transport in Slovakia:

> He has his own busses and great drivers. As I said, they only take four or five care workers at a time, so that they really have space. They take breaks, they don't drink alcohol, they don't smoke, and they are friendly drivers. So now it is they who bring the women from Slovakia to Switzerland and back. It took about two or three tries for me until I found my current solutions.

At the time of the interview (2014), Dominik still worked with the partner business in Germany. He still organised the journey with them for new care workers and for care arrangements with 'irregular' jobs, such as one-time stints, whereas for more experienced care workers he worked with the second transport business, the one based in Slovakia, to enable a change of care workers every 28 days.

Within one year, Mattea's agency has tried working with six different transport businesses in Slovakia. Mattea described her difficulties in finding the right collaboration in the following way:

> When they bring 10 women, they want 2000 Euros for going and coming back. So that is 100 Euros per woman, per journey. So, if they have to bring only four women and only earn 800 Euros, then suddenly they don't answer any more. They make sure that they only benefit. . . . They really screw you. If it is big business, they show up, if it is small business, and not much comes out of it, then suddenly they are nowhere to be found.

Having no experience in business development before she started the agency, let alone business experience in Slovakia, Mattea generally found it difficult to find collaborators that she could trust and develop her business with in the long run.

> You have to know the way they are there. They are envious people, and constantly scared that someone could screw them. They say it themselves. That is just the way it is. They have very little trust. And if they see that you have something, they want the same. And because of that, you always have to be cautious.

The collaboration between her agency and her transport partners shows how transnational business practices are based on and construct stereotypes along the lines of nationalities. Although it is not clear where these stereotypes come from – and beyond the scope of this thesis to analyse – the example of Mattea's difficulties in collaborating with transport partners indicates the importance of care agents' own social contacts and social relations in the building of the live-in care workers' transport infrastructure.

The firms that Mattea's care agency worked with used cars that could seat up to 14 people. The agency included a travel fee of 300 Euros in their all-inclusive package for care recipients. The care workers were not required to use their transport offer but could organise their journeys themselves if they wanted. However, they also had to pay for those journeys themselves if they do so. In the beginning, Mattea also tried to organise flights. However, after three care workers missed their flights at the airport in Budapest, Mattea decided not to organise flights anymore and only offer car journeys. As mentioned earlier, I had the chance to ride with her from Slovakia to Switzerland. She did not usually transport workers herself, but every once in a while, she would drive to Switzerland anyway. On that occasion, she took two care workers in her car, Emilia and Blanka. At the same time, two eight-seater cars were driving to Switzerland with care workers.

I perceived the journey as very tiring. We drove on the highway for hours. Between 3 a.m. and 6 a.m., I almost could not pull myself together. Emilia and

Blanka had already fallen asleep in the back, hours before I started to struggle to stay awake. I was glad at that moment for the darkness outside; it made me feel less guilty about closing my eyes. I was aware that Mattea's offer to ride with her to Switzerland was not entirely a selfless act; it was also pleasant for her to have someone to talk to and hence to keep her awake. And we had talked a lot, all four of us. Mattea also hoped that I could drive in between for a while, but I did not feel secure driving and politely explained so. It is a long way from Slovakia to Switzerland. Mattea explained that she had had a lot of trouble with drivers. She claimed that she had repeatedly told their transport partners that there had to be two drivers to ensure the safety of the care workers. She was happy that she had finally found a partner she could trust.

At around 6 a.m., we arrived at our first destination. It was the home of an elderly couple somewhere in eastern Switzerland. The village was quiet, and it was a surprisingly fresh morning. It felt a little chilly, not just outside but also inside in the small, old, single-family house. This is the place where Blanka was to take over the care of the elderly couple for the next month. She had been working there for a couple of months. The care worker she was replacing was preparing breakfast for the elderly couple, who, in their pyjamas, came to see what all the noise was about before going back to their rooms. We gathered around the kitchen table to warm up with some tea. I felt weary. It was difficult for me to imagine how Blanka was going to start her work the same day. Around half an hour later, another car with a man and seven women pulled in to the parking lot. Mattea had fixed a meeting with the driver (Figure 7.2). THE driver: there was only one.

Figure 7.2 Meeting in front of Blanka's workplace, picture by author, August 2014.

Conclusion: unequal power relations in the matching process and in the organisation of the journey

In sum, care agents construct the legitimacy of their agencies in two ways: 1) by building 'professional' recruitment and matching infrastructures and 'standardised' screenings; and 2) by stressing and mystifying their own recruitment and assessment experience and skills. However, although constructed as one, the key to a working care arrangement is not a mystery. It lies in the efforts and work of all participants to make this arrangement work, such as the ability of care workers to be ready and flexible and, as we will see in the next chapter, to adapt to a household and to create a caring relationship. Hence, it is also connected to care agents' and recruiters' practices in recruiting care workers they assess as flexible and adaptable. Furthermore, matching is not just interactive but also relational work. That is where its value comes from. Care agents have to think about how care workers can build a relationship with the care recipients and their family members. They do so either by building interactive relationships with care recipients and care workers or by creating a feeling of interaction between care recipients and care workers before a placement. This in turn affects the development of their own relations with care recipients as clients. Besides the fact that a successful match with a long-term arrangement reduces costs for a care agency, it establishes trust between clients and care agencies and eventually brings references and reputation.

As for care workers' journey to Switzerland, I show here that the use and development of transport infrastructures for care workers is deeply embedded in social relations. Care workers and care agencies have to think about how care workers reach their workplaces. Private transport services and public transport are the options offered to care workers. With the emergence of new employment opportunities for care workers from Eastern European countries in German-speaking countries, a range of private businesses emerged that specialise in the specific mobility patterns of live-in care workers. As Papadopoulos and Tsianos (2013, 191) remark, the circulation of knowledge and reciprocal relations involved when migration occurs involves a multiplication of informal economies that cover economic activities and services that are difficult for migrants to access. In this sense, the care agencies and transport businesses can be seen as an essential part of care workers' migration (Papadopoulos & Tsianos 2013). The journey can be less expensive and more convenient for both care workers and care agencies with the help of private transport businesses, as private drivers can bring the care workers directly from door to door. Being picked up at home and dropped off at the workplace can facilitate the organisation as well as the journey. However, the bus journeys are often long and tiring and, depending on the care agency and transport business, more or less uncomfortable. Whereas Papadopoulos and Tsianos (2013) only refer to informal economies, I have shown that the migration of live-in care workers brings to the fore a multiplication of both formal and informal economic activities that cannot be clearly distinguished but should rather be seen as a continuum.

Finally, I demonstrate that the internal dynamics of both the matching process and the organisation of the journey are characterised by inequality. The decision makers in the matching process are the care agents and the care recipients. The care workers themselves have relatively little power to decide in which households they will be placed. Depending on care workers' financial means, access to private transport, and agencies' travel organisation, care workers experience very different forms of travel in terms of duration, costs, and comfort. The costs and profits of the journeys are negotiated between care agencies, care recipients, drivers, and care workers, and it seems that care agents give care recipients' opinions more weight than care workers'. The more that care recipients are ready to pay for the transport, the more comfortably care workers can travel without having to pay extra out of their own pockets. Having provided insight into the steps between recruitment and actual employment, the question arises as to how care workers arrive and settle into a household. The next chapter is dedicated to the moment that care workers arrive in Switzerland and set foot in their new places of work.

8   Arriving at the household

When I first met Sara in the early morning in front of Blanka's place of work, she had been travelling for more than 25 hours. At the same time as I was on my way from Slovakia to Switzerland with care agent Mattea, she was riding in one of the other cars. As her and Emilia's place of work was along the way to Mattea's own home, Sara switched to our car after crossing the border to Switzerland and we continued our journey. While I estimated Emilia was older, perhaps around 50, Sara was rather young, around 30. It was both women's first stint in Switzerland as care workers. Whereas Emilia did not say much, Sara was chatty and made jokes. She spoke Swiss German with ease and seemed very cheery, almost carefree, which in that moment felt slightly awkward to me. I had a hard time understanding her apparent ease over beginning this new employment, for she was about to arrive at a complete stranger's home and live with him for the next two months. Later I would realise that both Sara and Emilia were more nervous than I had originally noticed.

By the time we arrived, the sun had come out, warming up our bodies in the fresh morning air. Sara and I exchanged contact information upon getting out of the car. I asked her if she wanted to meet for coffee within the next two weeks, which she happily agreed to. We rang the doorbell of the care recipient's home, but no one opened. 'That's weird', Mattea said and tried to call Mr Hofer, the care recipient. 'He should be expecting us. Let's just wait', she suggested. Around ten minutes later two men in sportswear were riding their bicycles down the street leading to the house where we were waiting. It was Mr Hofer together with an acquaintance. The latter introduced himself as the person who would sometimes help Mr Hofer because – as we learned – Mr Hofer was not supposed to be alone and go biking on his own. While walking to the house door, Sara said to me softly, so the others would not hear: 'Maybe this is his partner? I hope he is! I hope he is gay. That would be perfect'. She was clearly not fully comfortable with the thought of living alone with a man in the same household.

What happens after care workers arrive at the households in Switzerland? How do care workers settle into the household as a workplace? In what way does the matching process described earlier play a role in this process? This chapter gives insights into the moments after care workers arrive at the care recipients' doorsteps and the time in which it becomes clear whether and how a care arrangement functions. The first part describes how care workers settle into the new environment of work. The second part presents how live-in care work is built on affective and complex relationships between care recipients, their family members, care agents, care workers, and their family members and friends in their home communities.

Settling into and adapting to the new environment

The first job: a journey into the unknown?

Back in the car, after saying goodbye to Sara, Mattea told me that Mr Hofer's arrangement with Sara was a special case. In comparison to the majority of the elderly care recipients that require the services of the agency, he was young, in his mid-forties. Since he worked part time, she would have a lot of free time and hence less remuneration than other live-in carers placed by the agency. Instead of CHF 2,100, Sara had agreed to work for CHF 1,900 a month. Mattea had told me in a previous conversation how important it was for her that the care workers would know what to expect about their care recipients:

> When you inform them as well as possible, then you have less problems. We tell them everything in the beginning, and then they say yes or no. If they say yes, they already know exactly what they are getting into.

She was convinced that in comparison to other agencies, the care workers placed by her agency would be well informed before starting their journey. Well informed, for Mattea, meant that the care workers received a sort of résumé of the care recipient as well as an information sheet they had to sign, stating that they are not allowed to consume drugs, alcohol, and medical drugs without informing the agency. Hence, the given information refers to the care workers' activities within the household and to the care recipients' health and care needs.

I talked to Sara on Skype a couple of weeks later. The job turned out to be a terrible experience for her. She felt misinformed by the care agency about the care recipient's health conditions, her working conditions, and her residential status in Switzerland. The first two days had gone well. It was 1 August when Sara arrived, a national holiday in Switzerland. She and Mr Hofer had gone to a small gathering and watched the fireworks. After two days, however, Sara experienced how Mr Hofer had a sort of mental and physical seizure: 'It is a weird disease. He has seizures; his whole body is blocked. When it started, I first observed, I tried to understand what I had to do. But after four, five, days, it was too much'. Sara told me that she was required to be with him all the time, even when he was working. The only time she had some free time was when he had a meeting at work for two hours, during which she left his office and walked around in town for a little while. Immediately after the meeting, however, he called her, and Sara returned to his work place. Moreover, she explained, she did not know that she was required to drive a car. In one incident, while driving, he suddenly started to scream very loudly. Sara was shocked and felt scared. When she asked him what was wrong, he told her to just keep driving. 'That is not very tragic, I can live with that', she commented on his unexpected outbursts. However, she explained that Mr Hofer also required assistance at night to go to the bathroom. And during his seizures, he sometimes ran around in the apartment. 'And they did not inform me about that'. Moreover, Sara told me that Mr Hofer had complimented her and insinuated inadequate intentions. He told her that she was very pretty, offered to massage her, and asked if she wanted to join him in the shower. Sara called the agency and terminated her employment. When Sara wanted to collect her salary before leaving, Mr Hofer refused to pay her the originally agreed sum and paid her much less. Instead of waiting for the agency to organise her transport back to Slovakia, which was already paid for, she organised a train ride home on her own and left.

Sara's first stint in live-in care work was far from working out smoothly, and the situation in the household was not what she had expected. The working arrangement and the care recipient's health condition did not correspond to the recruiter's description she received before she started the employment. She found herself working non-stop and experienced inappropriate and uncomfortable situations with the care recipient. Moreover, she felt that she did not receive adequate support from the agency to deal with the situation. When she called the agency to ask for help, the recruiter asked Sara to stay until end of the month and to wait for the arrival of a new care worker. Instead of putting her well-being first and immediately organising her return upon her request, the care agency prioritised the organisation of a replacement for her to ensure the continuation of the care arrangement. Consequently, during the probation period, she terminated the employment early. Sara decided that she would never again work with the agency and perhaps not even work as a live-in care worker again in the future. Had the agency reacted in a more supportive way, she might have felt differently towards a future employment as live-in care worker.

Sara's example shows that the first stint can be, as care worker Elsa puts it, a 'pig in a poke' for care workers. Similar voices from both sides, care workers as well as care agents, emphasise that it would be impossible to know beforehand how the arrangement actually turns out. Berta, who works as a care worker in a direct and informal care arrangement in Germany, describes:

> How old they are, how much they weigh, what they can and cannot do, and what kind of disease they have; in 60% they tell the truth, in 40% they deny things or don't even say anything. One can only see on the spot what is going on in fact.

According to care agent Pascal, only after care workers have effectively started a job he can come to know if the care arrangement works. Even though Mattea had sounded quite convinced that the care workers would supposedly know what to expect at their first job, she would still 'not vouch for them', stating that because she had experienced many care arrangements going wrong, she would never be able to guarantee a smooth care arrangement. Therefore, how a care arrangement turns out is always uncertain, and the situation only becomes clear after the beginning of an employment.

Getting to know the ropes 'in a foreign place, in a foreign country, with a foreigner'

While Sara's situation was a particularly bad experience for her first job as a care worker, settling into live-in care work generally seems to be a difficult task for the care workers I talked to. Many of the interviewed care workers find the beginning especially challenging. 'I was crying in the first two weeks in Zurich', Anna, for example, described. As I will show in this section, the recruited women leave their homes and become not just care workers but also migrants. They find themselves in a completely new setting and in new circumstances. After arriving on the doorstep of a care recipient's household, care workers have to familiarise themselves with their new environment on a range of different levels, starting with everyday practicalities. Earlier I showed how care agents have different opinions on the importance of language skills in the recruitment of care workers. Some of the interviewed care agents did not think that German skills matter much for care workers to perform what they consider as 'good care work'. Pascal further explained:

> There are clients, with which you have to speak (. . .) German [well]. But there are also clients; it's maybe (. . .) an advantage if someone doesn't speak German well, because, if someone speaks German well, they will go crazy around that client. [Such as with] a person with dementia, who repeats himself all the time. If someone speaks (. . .) German, it will drive her crazy, because she thinks that she has to give an answer. Whereas someone who doesn't speak German very well, it goes in here (points to one ear) and gets out here (points to the other ear). When she says, 'cook or eat', then she will understand, or undress and sleep.

However, Elsa's experience points otherwise:

> You have to go shopping alone, to a pharmacy, you have to take her (care recipient) to the doctor. (. . .) When I arrived for the first time, I asked her on that very day, in the afternoon, if we could go together so that she could show me where the store is, and this or that. She said: You'll find it. So I just went and asked around.

Evidently, feeling secure and comfortable with a language helps navigating through a new environment. Therefore, knowing the spoken language plays a role for settling into not only the household but also the environment in which the household is set. Care workers have to cope with a new environment in a new village or city and in a new country.

Elsa was placed by Pascal's employment agency. Her care arrangement with the agency was organised by the daughter of the care recipient, who paid her salary as well as organised parts of her everyday activities of care work. She gave Elsa money for grocery shopping and to buy the care recipient's medicine. However, Elsa also faced dilemmas, in which she had to decide whether she would carry out tasks which the care recipient – who suffers from dementia – instructed her to do, knowing very well that she was not responsible for:

> She even sent me to the post-office. I paid bills, although I wasn't supposed to, because her daughter usually takes care of the bills. But she insisted very much. Well, you have to figure it out. In a foreign place, in a foreign country, with a foreigner. It is not an easy job indeed.

Another aspect in the everyday life of live-in care work concerns social norms on how to treat each other and whether care workers and care recipients and their family members understand and interpret these norms in similar ways: 'Evident things', as Krisztina puts it. It is important to her, for example, to never return to the family empty handed. She would also bring a small gift for the care recipient and sometimes for the family members, whether it would be jam, fruits, wine, or Hungarian chocolate. For her, gifting was a gesture that represents kindness. That gesture, Krisztina explains, was important for her to make everyone, including herself, feel good. Moreover, Krisztina shared that she would always ask the care recipient whether she wanted to eat certain things, for example a yoghurt, before she would actually eat it herself. 'These are evident things', she concluded. What Krisztina refers to as evident here are everyday understandings on reciprocity, how to treat each other, being considerate to each other, and how to communicate with each other.

However, care workers also deal with issues that are 'not so evident'. Krisztina arrived at her employer Pascal's place for the first time right around Christmas. She did not have any gifts ready, but she wanted to let the family know that she was undertaking great effort to contribute to a nice atmosphere on Christmas. As she perceived Switzerland as 'horribly expensive', she knew that she could not afford to buy gifts. Hence, she came up with the idea to create the Christmas decorations. She collected cones, waxed evergreen leaves, acorn, wood fibre, and other materials from the garden, and whenever she had some free time, she secretly worked on a table decoration and surprised the family with it. 'It was a marvellous, jolly thing', she recalled of her experience. Krisztina also asked the family about their Christmas traditions:

> At first, I didn't understand: I asked, who is going to cook for Christmas? They said, no one. All right, we'll see, I'll be quiet, we'll see, I'll just follow what is happening. By the way, when the big family is there, everybody does everything. They don't expect me to serve. Of course, when they wake up, I ask: do you want coffee? The same as in a family. I'm there, I come and go, stir around, cook, do you want a coffee? Come, I'll give you one. Do you

want another one? This is just natural. The point is, the family arrived; they brought meat for fondue. It was meat fondue.

During our interview, Krisztina talked at length about the kinds of meat the family served and how Swiss people would call chicken meat 'poulet' (the word derives from French) and how some terms are said differently in different cantons in Switzerland. Meat fondue is a popular dish in Switzerland for Christmas, and I have had it many times myself for Christmas. What this short story shows is how Krisztina gained knowledge about traditions, norms, and a place, not just in the household, but in the village, canton, or even the country as well. 'One pays attention, knows, sees things and learns by watching', she concluded.

Therefore, there are not only 'evident things', as Krisztina would put it, but also 'new things', which care workers encounter after arrival at the workplace. It is, however, not always clear to distinguish between the two, that is, to recognise which social norms and logics apply in different situations in order to navigate through everyday life. As Gutiérrez Rodriguez (2010, 146) argues, with an example of Latin-American migrant domestic workers working in German households, the encounters between employers and domestic workers 'bear the traces of transculturation, where two worlds meet that are geopolitically, economically and culturally separate'. Gutiérrez Rodriguez points to an important aspect of live-in care work: Care workers are not from the same place as care recipients. Live-in care workers often come from very different environments and financial circumstances than the care recipients. This situation, as I have argued earlier, is not a coincidence but is constructed as such and inherent to live-in care work. As a result, live-in care workers come from places where the same social customs and norms may not apply. Hence, care workers have to constantly learn and communicate in order to understand and 'get the ropes' of care work and everything that comes with it.

Consequently, care workers continuously learn how to perform care work for the elderly and develop while at work. Marina, for example, said:

> Although I have been practicing it for a long time now, till today I'm still learning German. When I talk to my old man [care recipient] – he is a very cultured, intelligent person – if I can't express myself, he helps.

Whether and how much care workers learn from the care recipients and family members, however, varies from case to case, as Elsa stated:

> One person is like this; the other is like that. Some are more silent; others are more talkative. Some need you to talk all day; others don't need it at all. This one girl, she told us she didn't learn anything out there [abroad/on assignment], because the old lady (*néni*) doesn't communicate at all. She only answers with yes or no. My Lucia [care recipient], where I work, she can talk 24 hours! I learnt a lot from her. But this all depends on the place, right.

Furthermore, care workers use the Internet to assist with communication issues, as Sandra, who was very proud of how she performed in her first job, put it: 'The Internet helped. Anything I didn't know how to say, I used the web translator to translate it. I made myself understood. There were no problems out there [while being abroad]'. While the Internet served as a useful tool to translate languages for Sandra, the Internet was also a big help for Marina to acquire knowledge about the care recipient's health condition and to feel more secure about how to take care of him. She stated:

> So I knew I wanted to do elderly care, and I knew they are sick people, but I didn't have a clue what nursing/caring [*ápolás*] was about. Although, I attended to my mother [*megápoltam*], she also had liver cancer, but some-how the situation was different. A stranger, the responsibility, and so on. So I felt that if I undertake this job, I owe a great responsibility, and I have to learn and check things (. . .). So (. . .) right away I started to put the diseases in Google search. (. . .) I started to acquire information. (. . .) It was my first stint, I had this laptop, and I thought I would need it to keep in contact with home, and when I experienced on the spot what kind of diseases they had, I immediately searched for it in Google.

Krisztina gives another example of how indispensable the Internet has become, when she proudly told me how she learned to cook excellent Indian food by 'googling' recipes on the Internet, when she was looking after an elderly Indian man. These examples show that care workers arriving in a new place, in a new country, to care for elderly care recipients have to navigate through everyday life both as migrants **and** as care workers. The intersection of the set of challenges that comes with each category brings new insecurities into their lives on many different levels. There is no general handbook for them to address these insecurities. Care workers have to draw on their own creative resources among others by researching questions online or in direct interaction with care recipients and their family members, in order to cope with their new roles.

We have to adapt to their habits

It might seem to be coincidence or just bad luck that a care arrangement does not turn out as expected for care recipients and care workers. But, as illustrated with Sara's case, there is much more behind the creation of a working relationship than simply luck. In Chapter 6, I note the importance for care agencies to have a pool of care workers to draw from in order to plan care workers' schedules and to exchange care workers if an arrangement does not work well. Interestingly, man-aging director Yvonne claims that changes occur much more often with outpatient care services (hourly care/mobile care) than with long-term live-in placements. The care agency she works for offers both services: hourly as well as live-in care. However, she finds it important in both forms of care arrangements to not 'throw in the towel', but to find ways to make a care arrangement work if difficulties arise between care workers and care recipients. Hence, how fast a care worker is replaced very much depends on how care agents react in challenging situations. Anthony, a care agent, for example, stresses:

> We had the case, where after three or four days I called [the care recipient's family] and they said it is impossible with this person, we need a new one. Then I said, please wait a little; this is a new situation for everyone. And then we agreed that I would call again after six days, that we would try it for ten days. After these six days they didn't want to let her [the care worker] go any-more. So really, it needs a certain amount of time to get used to the situation in the beginning, that is not always easy.

Hence, some care agents encourage care recipients and care workers to get used to each other in the beginning.

Care workers must accept and be accepted by the care recipients to make the arrangement work. This can take time, and it can require endurance and patience from both care workers and care recipients. As care agent Yvonne explicated: 'These are sick people in an extreme situation. Someone strange enters my intimate sphere, my apartment and on top of that, approaches my body. That means, it is a double, a triple break-in to my intimate sphere'. Care worker Elsa recalled her experience: 'It was very tough. It was tough until she accepted that a stranger lives there together with her. Just imagine. It can't be easy. Damn not. Nor is it easy for us'. The elderly person she cares for has dementia and often does not understand why Elsa is there with her in the first place. Clearly, it is easier for some care workers to settle into a household than others. Krisztina described her experience as a 'natural simplicity' with which she integrated into the family. After her arrival, she soon felt that it was a place where she would feel good. She appreciated how the family gave her a gift and a card for Christmas. She felt welcome and treated like a family member imme-diately. The degree to which this process of integration is 'natural and simple', however, very much depends on the care worker's ability to adapt, as implied by Sandra:

> Patience, of course, but she should endure the other's fancies as well, and she should be understanding, and I don't know. I think it's not like everyone is able to do that. There are people who are simply unsuitable for this work. One even has to feel like she fits in. Or to start with, [one has to make] an old person accept her. Just imagine! A stranger goes there, a complete stran-ger, s/he also has to be able to accept her. With me, it took two weeks until she accepted me, and until I finally felt okay myself, [until] I calmed down. I know by now [the] 'what' and 'how'. And getting used to each other. For that we have to accept the other [and] the way we are. Because someone who always thinks she is smarter, and you know, I can't explain this, she shouldn't even go. Because it's not about that. It's about being humble.

Talking about her experience in the household where she worked before her arrival at Pascal's family, Krisztina said:

> He accepted me, because, if someone gets to a strange place, a strange family, I have to adapt. Fantastic tolerance, fantastic adaptability is needed for this job, because the old person won't change habits. But I have to adapt to those habits.

She emphasised how important it is to also adapt to the eating habits of the elderly. In our conversation, she expressed pride that she was not picky with food and would eat anything. For her, knowing how to make the elderly feel comfortable is fundamental for care work. However, not everyone can adapt to the care recipient's habits as easily as Krisztina described. Elsa compared the live-in situation with regular employment, where one goes home after finishing work:

> Here at home [in Hungary], if we go to work, we have to adapt too. The problem is that there [in Switzerland] it is 24 hours a day. We eat together; we are 2 centimetres next to each other. If you eat little, they ask, why aren't you eating? Why don't you eat, what she eats? I can't imagine eating jam for breakfast, but she asks me why I don't eat jam. Minor things. And she asks about it every 5 minutes, until breakfast is over. And you should put on a good face.

Hence, whether a care worker is able to accept a care arrangement, and is accepted by the care recipient, is very much related to the care worker's attitude and understanding of care work. Elsa and Krisztina both seemed to be very conscientious about their care work and intermittently expressed their attitude that the care recipients' well-being should come first. Elsa, for example, usually waited until the care recipient was in bed before she used the Internet to contact her family. She talked to her family every evening on the Internet. However, she did not tell the care recipient about it and would murmur in a very low voice, so that she did not disturb her. In her view, she was not paid to 'hang on the Internet', but to keep the care recipient company so that she was not alone. Krisztina saw it similarly:

> Because when we started, Pascal [care agent] said, and first I didn't understand why he said that, that we don't go there to rearrange the apartment according to our own taste, and so on. But it's true. If she puts that thing there [on the table], it shall stay even if it is not a pretty sight since it doesn't belong there. So one shouldn't . . . One has to be humble. If she says it's black, although I know it's white, then it's still black. (. . .) Because it's her apartment, it's her life. I am only there to help and look after her. We don't go there to rearrange things and transform her at the age of 91, but to accept that she is the way she is.

What becomes clear in these examples is that in order to create a functioning care arrangement and to settle into the household, Krisztina and Elsa co-construct a need to be patient and humble, to accept that the care recipient's needs come first, and, hence, to adapt their own needs and preferences to the care recipient's habits. Likewise, Schilliger (2014) states that working in a private household implies that care workers have to deal with the habits and living style of the care recipient's family. She finds that it is mostly the care recipients' families that determine what is cooked, how it is cooked, where to buy groceries, which cleaning utilities to use, and when to eat. In some households, care workers do have certain autonomy and can carry out their work more autonomously. In other households, care workers have to adapt to the family's household rules, ideas, tastes, and habits (Schilliger 2014).

The care agents play a role insofar as they select care workers who they feel have a capacity to adapt to the care recipients' needs – as shown in Chapter 6 – and in that some of them prepare care workers accordingly, for example with cooking instructions. As care agent Dominik explained after having taken a phone call with a care worker during our interview:

> She is not complicated, and you can hear from her voice, she is totally motivated. We'll see. But actually, the patient is a difficult one. We'll see. He is also rather a calm type of person. I don't know if this one [care worker] is already a bit too much [not calm enough]. But, they [care workers] really know how they have to present themselves, how they have to restrain themselves [adapt to the care recipients].

He even constructed the association of professionalism of care workers with their ability to adapt to a household:

> But I also have to say, that I know these people in terms of mentality, at least those from Slovakia. Poland is similar. They are all very warm-hearted (*herzlich*). Of course there are some witches too, but most of them are warm-hearted (*herzlich*), courteous, over-motivated, so, they would always adapt. They are very professional too; they know how to behave on the spot with the care recipient. It is rare that we really have to replace someone.

Therefore, care workers are required to adapt to the habits of the elderly to create a well-functioning care arrangement. Whether a care arrangement functions depends on whether care workers are able to integrate into what Schilliger (2014) calls a power-pervaded social field to which they have to subordinate. Only if the care workers accept the rules of a live-in care arrangement and conform to the domestic setting can care workers develop the necessary emotional attachment and loyalty to the care recipients to make an arrangement work.

Affective and complex relationships in a live-in care arrangement

In addition to adapting to the elderly's needs and habits in a household, Schilliger (2014) finds that care workers suppress their own emotions and have to adapt their feelings to different situations. In this regard, care work has been conceptualised as interactive and emotional labour (see Hochschild 2003b; McDowell 2009). According to Hochschild (2003a [1983]; 1979), it is part of the work to manage one's own feelings, that is, to hide tiredness and subjective emotions in order to influence the well-being of the other person. Krisztina recalled her experience when she cared for an elderly couple:

> When the Papa was already so destitute, it happened one night that I woke up 11 times for him. He forgot he had been up. I took him out 11 times, to the toilet, with patience; I went back, put him in bed. In many cases, I wouldn't even wake up Mummy [the care recipient's wife]. He was impatient many times, once he said, 'I'll get the police to arrest you if you don't leave me alone', because he wasn't willing to dress clothes. And I usually asked Mummy what he said, I understood more or less, but they communicated with each other in French. The Mummy told me: 'I don't know, I didn't understand'. My poor, she didn't want to hurt me. But I never left with anger or hot-tempered. Every time I put him in bed, I caress his face or hand, so in a level, I was radiating a positively loaded energy towards him, which was good for him and good for me, too.

In this complex situation, Krisztina not only controlled her emotions but also implied awareness of the situation, in which the elderly woman apparently sometimes pretended to not understand what her husband said in French – a language that Krisztina did not speak – to protect her feelings. This act seems to have invoked feelings of appreciation and empathy in Krisztina for the elderly woman. For Gutiérrez Rodriguez (2010, 132), the concept of emotional labour refers to 'intention of the subject to be empathetic and attentive to others'. The performance is aimed at the well-being of a person. While recognising that the concept of emotional labour has its usefulness, she points out that there is a less rational aspect in the work of domestic and care workers and suggests analysing domestic and care work as affective labour. Affects, Gutiérrez Rodriguez (2010, 132) states, 'emerge in the coming together of bodily reactions and transmissions of feelings, leaving an imprint on a subject's body or environment and at the same time reflecting these sensations to other bodies'. Hence, affective labour incorporates the bodily and relational dimension of care work, taking into account that feelings emerge out of contact with others. The expression of affect, however, is 'not always intentional and clearly goal driven', but rather 'spontaneous corporeal reactions to our environment and encounters' (Gutiérrez Rodriguez 2010, 132).

'Somehow she becomes attached to you'

The concept of affective labour is useful for understanding unexpected and unpredictable developments in the emergence of feelings in the relationship between care workers and care recipients. Sandra, for example, stated:

> I was surprised myself. It would be an exaggeration to say I got to love the old lady (*néni*), but she just got close to us within a month. And I felt good, very good. I don't know. I'm not the type . . . Ok, this is private, but I'm not

the sentimental type of person who would pamper someone, but I got to love her. And I really did it with joy, taking care, you know. But really, at the age of 91, what can you do with her? Just caress her. And by the end she got to love me so much. So I don't know. Somehow she becomes attached to you. And if someone doesn't feel that, she shouldn't do this [job] at all. Because then she really only goes for the money.

As can be seen in the quote, Sandra did not expect to develop these strong feelings of love and joy in relation to the care recipient before she started her employment. Furthermore, she implied that the development of feelings and emotional attachment to the care recipients is necessary to build a good relationship and to perform care work. Similarly, care worker Krisztina related 'good' care work to the capacity of care workers to develop a personal attachment to the care recipients. She stated:

For an old and sick person it's very important that one puts all her heart into this. If it doesn't come from here inside, then the whole thing isn't worth anything. If you only pretend, if it's just a mask and you only do this out of necessity and for the money, then this job cannot be perfectly done.

These quotes imply that it is not possible to draw a sharp distinction between professionalised and commercialised service relations, on the one hand, and personal, private, and, according to Schilliger (2014, 243), supposedly more 'authentic' relations, on the other hand. For the care workers, their professional relationships with the care recipients have to become personal relations in order for the live-in care arrangement to work. By drawing on Vega Solís, Gutiérrez Rodriguez (2010, 128) stresses that domestic and care work requires attention and the skills to be attentive, and that 'the person who is cared for is not just an object of care'. Instead, being attentive means recognising the 'other person as a complex subject' and 'address[ing] the well-being of the other' (Gutiérrez Rodriguez 2010, 128).

Complex relationships in the household

The integration process into a household depends on the relation between not just care workers and care recipients but also all involved parties, such as the family members of the care recipients and co-care workers. For Krisztina, it is very important that the whole family appreciates her efforts to perform what she understands as good care work. Whether she feels good about herself and her work is very much related to how the family appreciates her and how they express acknowledgement of and care for her. Her motivation to work, to be willing to adapt to the care recipient's habits, to be able to make a care recipient and their family members feel good, and to prioritise the well-being of the care recipient is directly related to the affective feelings generated between all involved participants of the care arrangement.

Krisztina did not always have good experiences with family members of the care recipients. In one of her previous employments, she felt she was ill treated by the wife of the care recipient, who would monitor what she was eating. She felt that she did not get enough food. She started to work there after her daughter asked her for help as a 'switch partner'. 'I did not make a big fuss about it. I brought ham, sausage and cracklings for myself. My daughter did the same and we supplemented the food with that. If not, we would have perished'. She felt that she was being mistreated and considered more of a slave or maidservant by the care recipient's family, instead of as someone who was there to help improve the situation. Krisztina chose not to have an open confrontation with her employer and decided to just quit her employment. However, her daughter stayed for one more year, which was not easy for Krisztina:

> She was miserable and skinny like my little finger and the wife wouldn't let her home. She was sticking to her tooth and nail, because Lena [the daughter] is the type of person you can do anything to. (. . .) I told her, you're crazy if you stay there. But she stayed, and later came home, but never went to work abroad since then.

In this case, Krisztina developed negative feelings towards the employer, not just in relation with her own employment but also in relation to her daughter as a co-worker in the household.

When I visited Anna for the second time at her workplace in Zurich, she told me about her work situation at her other workplace (Anna worked for two employers and alternated between two households in Switzerland). She felt that it was becoming difficult for her to work there because of conflicts between the care recipient's family members, herself, and her co-worker. Her 'switch partner', Teresa, who is from the same village as Anna, had started to date the son of the care recipient, which apparently created tensions between the involved parties. The son had left his apartment and moved in with Teresa and his mother. During Anna's stints, the couple usually went to Hungary and lived at Teresa's place. Hence, Anna and Teresa usually did not see each other and did not talk to each other. The conflicts, in Anna's opinion, revolved around the fact that the family treated Anna and Teresa equally, but Teresa felt that she should be better treated for being the girlfriend of the care recipient's son. Moreover, Anna had a difficult relationship with the son. She did not like him and claimed he was mentally ill. Her relationship with the care recipient's two daughters was better. They, too, supposedly had difficulties with the son. According to Anna, they criticised that he lived in their mother's home and that there would be no food and other household supplies left when he and Teresa left for Hungary. It was a problem for everyone that he would not buy any groceries and household supplies, which annoyed the sisters, and thus Anna found herself in a working place with strained relations between the people around the care recipient.

Furthermore, Anna was sad about an e-mail from the daughter of the care recipient. She showed me the e-mail, which was sent to both Anna and Teresa. The letter instructed them to carry out a spring cleaning with detailed instructions on how to clean the house and a time plan on when and which rooms should be cleaned. Although Anna suspected that it was more directed towards Teresa than her, she still felt very disappointed about the e-mail. She had cleaned the whole house without being asked to do so before Christmas and felt that she generally kept the house very clean. She stressed that the only things she did not clean were the care recipient's son's things, as she did not want to be accused of snooping around in his private affairs. Anna felt that her work was not recognised and could not understand why. Moreover, it was difficult for her that the daughter did not talk directly to her but instead sent her an e-mail. The content of the e-mail was written meticulously. Although it was written in a polite way, it was clear that the writer had placed an order couched within a detailed idea of how the house should be cleaned. Furthermore, it was new that Anna found notes all around the house with instructions from Teresa on how things were supposed to be done. Anna was very angry with that and commented sarcastically: 'After six years I still don't know how to clean a table'. Finally, Anna expressed that it had become more difficult for her to develop sympathy with the family members because of a conflict between the siblings about the financial cost of care for their mother. Apparently, the husband of the care recipient had left CHF 100,000 to take care of her when he died. Anna suspected that the children did not expect that their mother would live so long. She had been taking care of the care recipient for six years. In her view, all three children earned well and would be financially well off. Thus, she told me, she had a hard time relating to them, explaining that if she were to find herself in such a situation, she would divide the costs of her mother's care arrangement between the children without even thinking about it. After all, she concluded, the '*Mutti*' (mother) had done a lot for her children.

The conflicts in and around the house were emotionally and physically strenuous for Anna, and she told me that she could not withstand the situation any longer. Thinking about her next stint back there made her feel queasy. Hence, she decided to look for new employment. However, she perceived it as a very difficult task to find employment with equal salary conditions, and so she kept working in the household with its complicated relationships for months while searching for a new job.

Although her relationship with the care recipient was very good, the workplace was imbued with feelings of stress and unhappiness. These feelings arose out of a perceived feeling of lack of appreciation by the care recipient's family members towards Anna and the complex and contentious relationships between the involved persons in the household. Anna's difficult situation demonstrates how 'affects are not free-floating energies' (Gutiérrez Rodriguez 2010, 132). The production and expression of affects emerge in a space that is marked by power relations. On the one hand, there are conflicts between family members, which affect Anna in her everyday work and life. On the other hand, she works in an environment where she pays attention and contributes to the well-being of the care recipient and her family members, but feels that this attention is insufficiently recognised. Consequently, the value of Anna's work and how she feels about it depends on her relationship with not only the care recipient but also the family members and her co-workers.

'There are hard days': building a transnational life with the help of ICT

Earlier I showed how live-in care arrangements in Switzerland are not meant for local care workers but are specifically tailored to temporary and repeating stints by migrant care workers (see also Chau, Pelzelmayer, & Schwiter 2018). For some of the live-in care workers I talked with, short-term circular migration can indeed be favourable, as it allows them to spend time with their family in their cities of permanent residence on a regular basis while earning money with (temporary) care work abroad (see also Marchetti 2013). For example, when Dania, a care worker from Poland who was initially placed by a care agency in Switzerland and was unemployed at the time of our conversation, was asked by her employer to stay in Switzerland and work permanently, she denied the request. For her, it was important to see her family regularly, and so she preferred to work one-month stints and take turns with another care worker.

Nevertheless, travelling long distances back and forth in short-term intervals for live-in care work can be onerous. Being away from home is not easy, as Elsa explained when I asked her how she handled going back and forth to Switzerland: 'It's bad. Long. One month is one month. We count [the days]. For everyone. We tick a day once it has passed'. Working with Pascal's care agency, she had been on three one-month stints as a live-in care worker in Switzerland. Although she appreciated the opportunity to improve her financial situation with care work, she felt sad about being away from home for work, as it impacted her social life and her relationship with her family. In previous chapters, I demonstrated that the model of live-in care offered by home care agencies in Switzerland is constructed as supposedly irreconcilable with having one's own family in spatial proximity and that the centre of care workers' lives is supposed to be with their own families in the recruitment country. As a consequence, and because live-in carers usually have to be on call around the clock, care workers lack possibilities to build social relations or gain social mobility outside of the care arrangement. Live-in care workers are often isolated in private households (Chau, Pelzelmayer, & Schwiter 2018). In this sense, live-in carers are encapsulated by the specific characteristics of live-in care work, which is marked by care dependency of the elderly and the private household as working space.

Live-in care work and circular migration affect care workers social networks and change their practices of building new or maintaining existing social relations. The Internet and social media became very important for the interviewed care workers to keep in contact with their friends and families. Especially when there are 'hard days, when it is simply necessary to talk to folks at home in some form', said Sandra, who chatted with her son and other family members two or three times every day. But it is also important to facilitate the management of everyday family life. During my visit at Anna's place of work, she expected a video chat call from her husband to inform her of how her son's exam went. She talked to him and her son everyday using the online video chat service Skype. The relationship, from a distance and with the help of Internet and communication technology, however, is very different than the relationship she used to have with her family, as Anna told me. She was deeply troubled about her changing relationship with her husband and son, as she felt that they were becoming more distant to her. As Anna worked in two different households in Switzerland in one-month periods, she hardly ever saw her family back home. For Christmas, in 2014, they had planned on having her husband and son visit her in Switzerland for a couple of days, since she could not leave work. However, they cancelled their trip. Anna felt deeply disappointed and expressed that she questioned their appreciation of all her hard work because of that. Consequently, whether Anna felt appreciation of her work was also related to her relationship to her own family back home.

Finally, the Internet can signify more than just a communication technology bridging geographical and physical distance to friends and family members. Working as a live-in means that there is much less private space to withdraw to than in other employment situations; even if they can retreat to their own rooms, many of the interviewed care workers hardly ever feel as if they are off work, as they still remain on call for care recipient's needs. The Internet can provide a form of private space to mentally unwind from care work, as Marina explains:

> For me, actually, it is privacy. I can retreat to my privacy. If there wasn't Internet, one would be restricted. And anyway, they say, being next to an ill person means it sucks out your energy. Here, if I just play, or write silly things, or post funny things, with that, I'm charging up, because I'm switching off from this world in here. Hungarians at home, sitting in front of Facebook every day, I cannot imagine why they need it. There they can go to see the neighbour, or meet friends, acquaintances. And married couples sending messages to each other on Facebook, that's ridiculous. But those who are isolated from people, for them, it's Facebook. It's a real community. No matter in which part of the globe.

Therefore, for Marina, the Internet provides space not only to build transnational ties and to connect to a transnational community but also to retreat to a private space and gather energy.

Conclusion: undervaluation of care

For women who have never worked as a care worker before, live-in care work can be a completely new situation on many levels. As they experience care work for the first time, they encounter the multiple facets of it, such as learning about diseases and developing personal and affective relations with the care recipients, their family members, and co-workers. The women become not only care workers after arrival at their workplace but also migrants. They have to familiarise themselves with a new environment. They have to constantly improve their language skills and learn about social norms, customs, and traditions in their place of work. They deal with their own emotions of being physically away from families and friends, build new relationships, and redefine relations with friends and families. Even more experienced care workers have to face new individual habits and personalities of care recipients every time they enter a new care arrangement. Moreover, relationships between not just care recipient and care workers, but also all involved parties, are not static but change with time. As the relationships change, new issues and conflicts can arise with which care workers have to cope. These affective developments are directly connected to their working conditions and the way they feel about their care work.

The key issue for a well-functioning care arrangement is whether the care workers are able to adapt to the care recipients' habits and to develop affective relationships in the household that culminate in mutual acceptance. The findings show that it takes a certain period in the beginning, during which the involved parties have to get used to the new situation and accept each other. The household is a distinctive type of workplace, as it is also the employer's living space. As McDowell (2009, 93) states, care work in a private household is a 'particular form of embodied interactive work in which guilt, ambivalence, love, trust and obligation are all part of the social relations involved in the exchange of care for wages'. Care work is affective work, and its value lies in its relational character with not just the care recipients but all involved parties of the care arrangement. However, it is a value that is never 'fully reverted' back to the care workers but flows into 'the individual reproduction of the household' of the care recipients (Gutiérrez Rodriguez 2010, 141). It includes the affective work of building new social relations between care workers, the care recipients' family members, and the care agents. Furthermore, care workers have to redefine their relationships with their own family members, usually with the help of ICT. Therefore, the household is not merely a space in an apartment but instead consists of a web of ever-changing social and socio-technical relations embedded within the larger contexts of how care work is understood and valuated in society.

The many experiences of feeling unappreciated described previously are not coincidental or individual incidents but tie into larger contexts of inequalities in care work, as described in Chapters 1 and 2. Conflicts around financial costs of care, undervaluation of care, hierarchies of who has to subordinate to whose habits in a household, and the unequal power relations that characterise the nature of care work performed by migrant workers are a direct result of what Nancy Fraser has problematised as 'crisis of care': the fact that capitalist societies free-rides on activities of care work and social reproduction 'although it accords them no monetized value and treats them as if they were free' (Fraser 2016, 101).

9   Care of care workers

'Can I have your number and call you, Mattea? In case I have a question?' Sara asked the care agent, while waiting for the care recipient to open the door upon arrival in Switzerland. Mattea did not give her phone number but instead told her to contact Leila, the woman who had recruited her in Slovakia, in case she had any questions. As seen in the previous chapter, Sara ended up having many questions and requiring support from the agency. What kind of support can care workers rely on when dealing with problematic issues with care recipients? In what ways do care workers draw on their own resources and social networks, not just to cope with care work but also to enable their circular movements and stabilise their care arrangements in general?

The role of care agents on care workers' well-being and working conditions

Care workers' well-being and working conditions start with the agencies' pre-arrangements with the households. Some of the care agencies stipulate certain minimum requirements in a household before care recipients and their family members can start a care arrangement. Usually, the family members of the care recipient have to organise an Internet connection, prepare a room and a bed for the care worker, and, according to care agent Dominik, 'ideally already provide some groceries before she arrives'. However, even though some of the care agencies explicitly require these conditions, I have come across stories where care recipients and care agencies did not ensure them. Berta, who was placed by one of the interviewed care agencies before she found direct employment, did not have a private bedroom:

> Imagine, it was a living room. And I had to like . . . (makes an unfolding gesture) sleep in a bed in the living room, and his cat was around who didn't let me sleep, and neither did the old man (*bácsi*), so I finally said, I'd rather not do this work anymore. Oh and from the bedroom he would pass by me at night to go to the toilet. So I was there for three weeks. I announced to the agency that I can't do this anymore. Even if I was given 3,000 Euros, not even then! I can't do this. I must sleep at night.

In this case, the agency clearly prioritised the care recipient's wish for a live-in care arrangement over the care worker's well-being.

Understandings of care

Once the care workers have arrived at their places of work, the support they receive from care agencies varies greatly. How care workers are supported partially depends on the care agents' and recruiters' experiences and education in the health or care sector and their understanding of the nature of care work. For example, Yvonne, managing director of a placement agency with years of experience in a private health clinic, acknowledged that care recipients' diseases such as Alzheimer's require adequate skills and experience and that it impacts care workers' concrete working conditions. In contrast, care agent Pascal, who had never worked in the health and care sector before starting his agency, focused only on the basic everyday needs of a care recipient in the household and moulded care workers into 'helpers' that are there mainly to assist the care recipients in the households. Correspondingly, before care workers' first stints, Pascal instructed them about everyday practicalities, such as how to dispose of garbage, and on cooking customs in Switzerland, in the form of an information sheet:

> They have to read it, so they know what it is about. We're continuously developing this sheet now, since one year. It's about aspects such as, how to run a household, how, what, where. Then cooking. Simple things, such as garbage, how that works here. Grocery shopping, billing, that they don't mix the money with their private things. Then appearance and clothes. Otherwise, some women are going to run around in a training suit from morning until evening or she'll never get dressed properly. That doesn't leave a good impression.

The quote shows that the information sheet by Pascal is clearly not just meant as a support for care workers in the household but also as instruction on how they should appear physically. The care agents' attitude and approach to care work can be crucial for care workers' working conditions, for example, if family members underestimate the care situation. Yvonne explained:

> His wife said, I am here too [to support the care recipient and care worker], I sleep next to him at night. The job is easy (*Schoggijob*). Then I asked for a health report of that man. She didn't know that I went to medical school. I looked at that report and, that is not an easy job. That man is severely ill. And she [the wife] was emotionally exhausted. If you have to take care of the wife too, then it is a difficult situation. When I encounter these complex situations, when I realize that there are family members who are difficult, then I don't send anyone to this place.

In contrast to the agency that care worker Berta was placed with, Yvonne seemed to take a clearer stance on the protection of care workers' working conditions.

Similarly, care agent Dora, who is a trained nurse, stressed that it would be natural for her that care workers could consult her for professional advice in terms of care work:

> Today a care worker called me spontaneously and I told her to pass by my place. She needs advice, because, initially it was actually the woman (care recipient) who was the patient. Because of her we started that care arrangement, but now her husband's support is diminishing. Now she asks me if she can come ask me for advice, in her language [Slovakian], of course that is very helpful for her. She has a basic medical course, because these patients are not difficult in terms of medical care, so that was enough. So of course I am ready to give her advice, half an hour or one hour. That is not a problem. I can tell her, how things are done here, which medications are needed, and that she can ask the doctor. So that she is more confident. And she can ask me about special terms in her language. She speaks German well, but some medical terms are difficult. Normally, if everything goes well, I am only in contact with the family sporadically. But I am still, even after the placement, the primary contact person, for example in case of sickness or whatever happens. And death [is] (. . .) part of life too, so dissolving contracts and things like that . . . That needs consultancy too.

For Dora, it was important that the chemistry is right between not only care recipient and care worker but also herself and the care worker. She only recruits care workers with whom she feels she can build a good relationship, as 'in the end, she makes my mark, so if she does bad work, it means that I placed a bad care worker'. The relation between employer, care agent, and care worker here is very different from what care worker Sara experienced, as she received no support with the excessive demands she was exposed to and was asked by the recruiter to stay until the end of the month so that they could organise a replacement after she turned to the agency for help and support.

Moreover, no one took care of Sara's administrative work for a formalised employment. Hence, Sara was working in an informal arrangement, which she only realised when she had already started to work. When she confronted the care recipient and asked him when he would register her and take care of the formal administrative work, he only explained that with the previous care worker, who was from Serbia, he had also not registered her, and that she too had resided in Switzerland as a tourist. Before Sara started her employment, she had even asked the recruiter whether she had to organise any permission to work and register. The agency told her that she did not have to organise anything and that it would be taken care of. 'I got scared. What if I get into a police control and I don't have any permission to work? Then maybe I have to go to jail?' she told me. Sara was thus stressed on multiple levels; in addition to coping with the care recipient's mental episodes, she felt scared and was worried about working in an informal employment in Switzerland.

It seems that the agency Sara works with would not notice any problems unless the care workers or care recipients contact them via phone. 'The women know that if something is not okay, they have to call. We don't go there. We are not there. They have to tell us if something is wrong, and then we will take care of it', said care agent Mattea. Mattea has no background or personal experience with care work but is a lateral entrant in the live-in care domain. In order to compensate for that knowledge, she hired recruiters as well as sales employees that had previously worked as live-in carers and hence would be familiar with the situation and possible problems for the care workers. Yet, Sara did not receive any professional advice on health care when she felt stressed. The agency's support consisted only of the suggestion of providing a replacement.

Understandings of space

How care agents support care workers depends also on their understanding of space, that is, whether they perceive the household as a bounded space or consider its wider environment. For Pascal, for example – as already implicated previously in his statement on care workers' German skills – care workers only have to be familiar with social norms and practices around the household. When talking to some of the care workers that work with his agency, I realised that they think they are only allowed to work for three months in Switzerland by law. It seems that they were not informed that the three months per year rule was introduced by Pascal as part of his business model and is not stipulated by admission regulations. In this sense, care workers working with Pascal are, to a certain extent, more encapsulated than care workers with some of the other agencies. This was reflected in Pascal's answer when I asked him whether he had introduced any clauses that care workers are not allowed to work with another care agency or to continue the employment without the agency (such as stipulated by some of the other agencies):

> No, no. (. . .) They don't know anything in terms of organisation. They are picked up by bus, and dropped off. Then they are 24/7 around that person [care recipient]. Then they are picked up again and are glad when they are back home.

In contrast to Pascal's placement practice, care agent Michael had a very different view of the transnational placement of migrant workers. In the following quote, he referred to his experience in the placement of migrant workers in the health sector in general, but claimed that support by the care agency was equally important for live-in care workers:

> For me, integration is the keyword. Without integration, they don't have a chance with the placement [and being] from other countries. You have to live integration, because, **we don't place a glass,** but a human being. And for human beings I have a responsibility, if I go get them from there. We have nurses, 95% of them are still at the same place, where I placed them. Those people are cared for, integrated, integrated into the culture.

If that does not happen, it won't work. That's why we are successful with
our placements. That is crucial. (. . .) That's the difference from so-called
modern slave trade, the integration of foreign people from foreign cultures
in Switzerland.

Hence, Michael took into account intercultural aspects, specifically that care
workers' lives are not confined to the space of a household. However, when
I asked Michael what he meant by integration, he only emphasised the importance of there being a contact person in Switzerland for care workers, so that they
feel cared for and the workers have basic German knowledge. Thus, it is not clear
whether the care workers working with his care agency are better supported than,
for example, the care workers working with Pascal.

Some of the interviewed care agents were very much aware of the fact that
many care workers feel socially isolated at their places of work. Many of the care
agents stressed that care workers needed to maintain strong connections to their
own households and hometowns. Therefore, it seems they are aware that in order
for the care arrangement to work, live-in carers need to be connected to their families and friends back home in their everyday lives. Accordingly, the care agencies
make sure that there is an Internet connection. Pascal even organised one of the
care workers, Krisztina, as some sort of helpline for other care workers and works
with the local village German teacher, Irina, as a coordinator. During the time
back home in Hungary, the care workers are supposed to regularly attend German
classes with the teacher. Krisztina encourages the other care workers to contact
her via Skype: 'I tell them how I do things, I give them advise on how and what to
do, when they [care recipients] are not willing to take the medicine, not willing to
eat or drink'. Krisztina is not remunerated for her support service. 'I do it, because
I would do it anyway', since all the care workers that are placed with Pascal's
agency are her immediate acquaintances or from her community back home. She
seemed very proud of her role as a consultant and explained that she was one of
the care workers with the most experience. When I asked her how often she was
contacted, she explained,

> it happens that over one, two, three days, no one calls. And then it happens
> that I am on Skype the whole day. And then I tell the mummy [care recipient],
> mummy one minute, I'll come soon, but this is important, because someone
> new has just started to work, or whatever.

The care recipient here is the mother of care agent Pascal, and hence Krisztina
does not feel conflicted between her own care work and supporting other care
workers placed by Pascal.

I could tell that Krisztina felt well supported by Pascal:

> He is the stable background behind the women. He lays down stone hard
> rules concerning the women, but also what can be expected from the fam-
> ily, and the duties the family has to fulfil. He visits the women regularly, he

keeps regular contact with the family, and in case anything happens, he does take sides.

Although Pascal applies the same business model as Mattea's agency of constantly changing care workers in one household after three months, he seems to provide more support to the care workers than Mattea's agency. The reason for that could be that he is much more embedded in the care workers' community than Mattea, who recruits not just from one village and the surrounding area like Pascal, but from all of Slovakia and Hungary. Care worker Elsa expressed appreciation for Pascal's close relationship with her community by comparing her situation with the situations of care workers in independent and direct employment or placements by more anonymous agencies:

> In fact, I also came here [to work with Pascal's agency] because a lot of women [*asszony*] go out [go work abroad]. But it's this safety that there's a person out there, do you get it? Someone who organises, and you know where you're going, and if you have a problem, then . . . So you're not going in to the unknown [*vakvilágba*, literally in blind world]. Especially if you don't speak German.

It seems that the care workers exchange stories and experiences of how care workers are supported and how women in independent employment relations have to face difficulties by themselves. Care worker Elsa, for example, heard the following:

> Someone [that was not working with Pascal's agency] got so scared that, she was standing outside for two days, waiting for her acquaintance to take her home. [She waited outside,] because she was so scared that the old man would attack her at night.

Whereas with Pascal's agency, she would be able to contact him for support if needed and not feel alone in the situation:

> There was a lady, she had some health problem[s]. And she called either Irina [the German teacher] or Pascal, and they immediately solved everything. It's not like you are grinding yourself like, oh my God, what should I do now! (. . .) Pascal visits [the village] almost every three months. So we have a very close personal relationship.

Therefore, it seems that Pascal's involvement in the local community played a role in Elsa's choice to work with Pascal. Krisztina also expressed that she trusts Pascal, to the extent that she had not even read her own employment contract. Moreover, Pascal stated, the women from the same village would often know information or if something happened in relation to a live-in care worker faster than the care agent himself. The example of Pascal's integration in the care

workers' local community shows how formal and informal, private and professional relations are very much entangled.

Drawing on care workers' own resources back home

Care workers' own health and outsourcing of costs of care to recruitment countries

Earlier I described how some of the interviewed care workers postpone or set
aside their own needs by putting the well-being of care recipients first and by
adapting to their habits. In addition, care workers also draw on their own social
and physical resources to perform care work. When Krisztina's daughter visited
Krisztina at the care recipient's house in Germany together with her boyfriend and
stayed for a month, Krisztina asked her daughter's boyfriend to help her lift the
elderly man and take him to the bathroom. The care recipient's health and mobility
had deteriorated rapidly and she was not able to lift him by herself anymore. Live-
in care work with the elderly is physical work and, depending on the care recipient's health status, it is often very hard work. It can affect a care worker's health
both physically and psychologically. 'Not in every case, there are easier places',
Sybille – who works in a direct employment – differentiated, 'but last time I came
home so exhausted, because the elderly fell and I did not dare to sleep at night,
because I had to pay attention, whether he attempted to leave his crib bed'. She has
pain in her hip and waist from helping the elderly move around and had to see a
physiotherapist back in her hometown in order to get ready again for her next stint.

   Anna told me that after six years of live-in care work, she was struggling with
pain in her knees and hips and had to undergo surgery back home in Hungary.
Moreover, she felt tired and emotionally strained. When I visited her at her place
of work, the care recipient would intermittently scream very loudly and Anna had
to go calm her down. The neighbours had apparently complained about the noise
and she did not dare to leave the house during her free time – two hours every
afternoon from 4 p.m. to 6 p.m. – as she was worried that the care recipient would
scream during the time she was away. Marina, another care worker, was very concerned about her own health while taking care of an elderly man with a bacterial
disease. She suspected that she had contracted the disease from him twice. As she
was very worried about the infection, she scheduled an appointment with a doctor
in the city of her place of work. In our conversation, she stressed that she had not
been to a doctor outside of Hungary for six years, as she would usually postpone a
consultation until she returned to Hungary. In addition to her own health, Marina
was also worried with the development of the care recipient's health situation. The
elderly individual had cirrhosis, a condition in which the liver does not function
properly, and Marina expected that his health would deteriorate:

> So it will be hard, but I feel that we are so human, we are so close to each
> other, and the family is also very good, so it's a good feeling to be here. But

there's a limit when one says, my condition, my health. (. . .) And it is not good for me if I get sick. Actually, I'm not protected from things because I don't receive sufficient information. For example, in the case of an infection, there are rules that prescribe the usage of certain tools. If I only use disposable rubber gloves, I am not really protected. I really like my job, but the line will be when he gets into the final stage of his disease, which I won't know because I didn't study this profession, I am not a nurse, I only do care work. It means that I take care of him, I help with his bodily hygiene, and I do the work necessary around him. If it [the job to care for the elderly] exceeds my abilities and I can't do it and I'm not even supposed to do it, then it is obvious that I have to say goodbye, at any cost. I would even lie if I have to, because unfortunately, the elderly would not understand it for sure and I wouldn't want to hurt him.

The quote demonstrates that Marina is worried that she will be overstrained because she does not feel adequately trained to deal with the care recipient's disease, especially when his situation deteriorates. The long-term consequence for Marina is that she can hardly sleep, not even when she is back home in Hungary, as she is not used to deep sleep anymore. It seems that Sybille, Anna, and Krisztina, and many other care workers, have gotten used to sleeping less, to sleeping lightly, and to sleeping with interruptions (Chau, Pelzelmayer, & Schwiter 2018). Hence, live-in care work can take a toll on care workers' on health and well-being, as they tap into their own physical resources to perform care work. Often, they wait until they are back in their home country to see a doctor or to go to a hospital, as many do not have health insurance in Switzerland, where health care costs are much higher than in Eastern European countries. Hence, the costs of their own care remain in the recruitment countries.

Social support infrastructure back home

Care workers support each other not only in case of difficulties with care work but also at home. As some are abroad and some are home, social media platforms such as Facebook can be used to support each other, as the following extract of the group conversation in Kicsifalu shows:

NORA: We have our own group. When we are home, we usually get together. This is very good. The name is *Valamikor más levet szívunk* (if we are abroad, we breathe a different air). Yes, we support each other, if we are abroad, always support each other, and at home as well. We exchange recipes, and help, I have this problem, that problem. Because half of the group is always abroad and half of the group is home, so we can't meet everyone. We are 18 members.

ANNA: There are a lot of women working in this village. I don't know how many, but there are a lot. Yes, we women support the family.

KATA: Yes, it is a small group, but we stick together. We are like a small family. We help each other carrying the food and we cook together. It doesn't have to be necessarily in this pub; sometimes we meet in our houses. We bake cakes.

NORA: When my Dad was sick, it was just me and my grandma. I ordered food for eating. I just posted on this Facebook group that we need help. And they came.

Going abroad for work, especially for care work, seems to be a common and well-known phenomenon in Kicsifalu, as I showed earlier in Chapter 5. In order to deal with the absence of the many women working as live-in care workers in Germany, Austria, and Switzerland, and the corresponding care gap in the village, the care workers and their friends and families back home take care of each other. During my visit in Kicsifalu, I learned that many households of care workers, such as Anna's husband and son, make use of a service that provides lunch and dinner for the elderly. It is organised by the community and delivered daily to their house on bicycles. The family makes use of this support, even on the few days when Anna is home. For lunch and dinner at her house, she complemented her own cooking with the food delivery, making it easier for her to provide lunch for the whole family as well as me and my field work collaborator Katalin during our stay. Therefore, to fill in the care gaps of the many migrant women working abroad, families and friends have to reorganise care in the home communities, leading to the emergence of alternative infrastructures of support.

Conclusion: the importance of social support infrastructures

Care workers share knowledge about their different experiences and how to support each other. In this sense, labour migration for live-in care has developed its own dynamic, with new practices such as online video chatting to maintain relationships with family, support in Facebook groups, and new community structures such as the provision of food to the left-behind elderly in the care workers' home villages. Furthermore, care agents can support care workers in difficult situations and provide a certain sense of security. Depending on the agency and their business practices, they play different roles in this last step of placement. In Chapter 7, I showed that it seems to be important for most, if not all, care agents to create a well-functioning care arrangement with their matching activities, despite the fact that care workers can simply be replaced if an arrangement does not work well. I have argued that the underlying reasons for this business practice lie in the reduction of costs, the importance of maintaining good relations to clients, and the creation of reputation for care agencies. I have also showed that care workers are not powerless but are in control of their migration. They can end their employment and leave at any time. In this sense, care agents provide care and support for care workers in order to encourage and stabilise long-lasting and well-functioning care arrangements. For this purpose, some of the care agents advise both care recipients and care workers to give time for the relationship between care recipients and care workers to develop. Moreover, many care agents make sure that care workers have Internet access and, hence, the means to stay in contact with their own family members through online communication programs. Therefore, care agencies play an important role in care workers' support structures and the production of infrastructures that facilitate functioning care arrangements.

However, the importance of creating a long-lasting arrangement and the effort put into it for the care agencies differ, I would argue, according to the main business models of the agencies. My findings show that the degree of support for individual care workers in the household depends on the importance that care agencies attach to the creation of a long-lasting arrangement. Employment agencies, whose main care arrangement model is to continuously place new care workers into one household, such as the ones that placed Sara and Berta, seem to provide much less support for individual care workers if they encounter difficulties with care work as well as with administrative work. Their business model does not rely on the creation of long-lasting care arrangements but consists of the provision of low-priced offers, keeping costs low through a frequent change of care workers in one household. In contrast, employment and placement agencies whose main business model is based on long-term care arrangements with permanent care workers, usually two in one household, seem to provide more professional support in relation to care work. Moreover, the findings point to a tendency that care agents with a strong background in the health sector attach more importance to the well-being of care workers and to the provision of adequate and professional support. Although Pascal's employment agency also follows the model of changing care workers every three months, and although he did not have any experience in the health and care sector, it seems that he nevertheless attaches great importance to the support (as well as control) of care workers by making sure that the care workers supported each other. His agency is strongly involved in the local network of care workers. It could be argued that Pascal cannot afford to not support care workers in difficult situations and to just replace them with a new care worker as other employment agencies do, as the news would spread quickly in the care workers' community. This would lead to a bad reputation for his agency in the village. In this sense, Pascal's integration in the care workers' communities could also function to monitor the care agency's activities and responsibilities.

Care agencies can try to increase the chances of a care arrangement functioning well by supporting care workers in difficult situations. Therefore, whether care workers are able to create caring relationships with care recipients and their family members and to invest in these relationships in the long run depends on their wider support infrastructures or lack thereof. Care workers' care for each other and their families, and care agencies' support for care workers, is an essential part of the migration infrastructure of individual live-in care workers. On the one hand, the support infrastructures are relevant to well-functioning care arrangements within the households. On the other hand, they serve as an anchor for care workers' frequent travel between their own homes and their places of work and their intermittent physical absence at home. Practices such as online video chatting to maintain relationships with family, the sharing of experiences through groups on social media, and even the provision of food to the left-behind elderly in the carers' home villages are important moorings for care workers' capacity to migrate for care work.

10 The middle space of migration and time for a care revolution

In this book, I addressed the black box of live-in care labour migration in Switzerland. I asked how migration, the act of movement from one place to another, is facilitated. I argued that a focus on actors and practices that enable the movement of migrant workers is relevant to understanding how labour migration is controlled. To find answers, I traced the mobility of live-in care workers and the practices and connections that are crucial for their mobility and migration.

Empirically, I delineated the journey of migrant care workers to live-in care work. I shed light on four phases in this journey: 1) how care workers access live-in care work and how care agencies recruit care workers; 2) the process by which care workers are matched to care recipients; 3) how care workers travel to their workplaces; and 4) the integration of care workers into their workplaces. I note at this point that care workers' journey of migration does not end with the establishment of a well-functioning care arrangement. Rather than a linear process with starting and ending points, it is a circular process. Although for some care workers a placement might be a one-time experience, for others it consists of a cycle in which they travel back and forth, ending employment relations and starting work in new households, sometimes with the help of the same agencies and other times with new agencies.

Conceptually, I paid attention to the migration infrastructures that facilitate live-in care labour migration, the 'institutions, networks, and people' (Lindquist, Xiang, & Yeoh 2012, 9), and technologies and material means that facilitate movement from one point to another. From this view, care workers' placements in Swiss households and the speed with which care arrangements can be organised depend on existing migration infrastructures. In particular, I focused on the 'middle space of migration': the role of agents in the migration of live-in care workers (Lindquist, Xiang, & Yeoh 2012, 11). I find that home care agencies play a crucial role in enabling live-in care labour migration. Firstly, their all-inclusive home care offers enable them to play an important role in making live-in care accessible and socially acceptable. Secondly, they can serve as an entry point for care workers' access to live-in care work and as a gateway to further contacts and informal networks that can lead to direct employment in private households. They facilitate the first step in migrant live-in care work by providing new channels for prospective care workers to find employment. They create new platforms for live-in care workers to access employment via their webpages, collaborate with recruitment agencies in recruitment countries, and tap into care workers' social networks and communities. They build infrastructures through which care recipients and their family members find care workers and thus facilitate the placement of care workers in Swiss households. Thirdly, by collaborating with transport businesses in recruitment countries or by setting up their own transport organisation, care agencies also play a crucial role in the development of new infrastructures to organise care workers' journeys and to bring them from their own homes to their workplaces. Finally, care agencies play an important role in care workers' support structures and the production of infrastructures that facilitate functioning care arrangements. In sum, they are key drivers in the production of a growing migration infrastructure that is geared to the circular migration pattern of live-in care workers. What is the outcome of this infrastructure? How does it produce and mobilise migrant subjects and, hence, reorganise social structures? The next salient point to be considered is the form of migration produced by the migration infrastructure.

Just-in-time stints and circular migration

Home care agencies offering all-inclusive home care play a key role in shaping live-in care by managing the spatial and temporal aspects of care workers' geographical mobility: that is, the (im)mobilities required to enable a live-in care arrangement. Care agencies produce a particular system of mobilities, and this shapes live-in care migration as a form of movement characterised by repeated short-term and 'just-in-time' assignments and frequent changes of households. The care agencies require a relatively high mobility of care workers: They should be able to start a job on short notice and travel back and forth between their own homes and their workplaces. The greater the capacity for mobility that prospective care workers possess, such as being able to leave at short notice and not having to take care of small children at home, the more access to live-in care work they have and the more likely they are to be recruited and placed in a household. Hence, caring for care recipients in their own homes is only made feasible by care workers' readiness to be mobile. The ability to travel back and forth was very much desired by some of the care workers I interviewed, as it allowed them to spend time with their families while earning money abroad. Nonetheless, it can produce difficult transnational living situations, place strains on care workers' relationships with members of their families, and lead to sadness while abroad. Moreover, care agencies offer all-inclusive home care packages with salaries that, while affordable to care recipients and their families, are not enough for care workers to live in Switzerland. This produces a degree of isolation for the carers at their workplaces; in other words, it immobilises them in the households.

Furthermore, I showed that care workers' comfort in their journeys depends not only on their own financial means and access to various transport services but also on whether care recipients are willing to cover their travel costs and on care agents' organisation of travel forms. What becomes apparent is the asymmetry in the triangular relationship between care recipients, care agencies, and care workers. The main people making decisions in the matching process are usually the care agents and the care recipients. The care workers have relatively little control in comparison; after being selected, they are usually only left with the choice of whether to accept or decline whatever employment is offered to them. In cases where carers are asked to begin a placement as fast as possible, those that are flexible enough to accept the conditions and start work the next day have an advantage over those who need time to organise their journey. Care agents try to increase a flexible disposition by creating pools of workers. Speed matters in the matching process. The higher the readiness of care workers, the faster they can be placed, the more care recipients and care agents benefit. Hence, those that are more flexible can actively weaken the opportunities of care workers that are less flexible. As shown, the latter are usually those with families at home in need of care. Consequently, it is not just other care workers that can be affected by this requirement for readiness but also the families of the carers themselves, who have to support this flexibility and organise their everyday lives accordingly. Practices such as online video chatting to maintain relationships with family, the sharing of experiences through Facebook groups, and even the provision of food to the left-behind elderly individuals in the carers' home villages are important moorings for care workers' capacity for mobility.

In sum, live-in care arrangements are constituted by a set of (im)mobilities embodied by care workers' capacities for mobility, their repeated movements between their homes and places of work, and their isolation once they have started work in private households. The mobility of care recipients' family members is anchored in care workers' readiness to be mobile and in their immobility once they arrive at the households. However, care workers' immobility in a household should not be seen as an absolute mooring but as a relative immobility. It is fluid in the sense that care workers do not stay in a household forever. On the contrary, they leave the household after a period of between two weeks and three months, and they move back and forth between Swiss households and their own homes in the recruitment countries. The carers' geographical mobility is in turn enabled by the care agencies' placement practices and growing infrastructures specialised in their movements, such as transport businesses, which serve as moorings.

This specific form of movement for live-in care work is not a coincidence, but rather the result of an interplay between the profit-maximising business practices of agencies and migration and labour regulations. It is specifically constructed as just-in-time and point-to-point migration to serve home care agencies' package offers of live-in care services. Care agencies play a crucial role in producing conditions that enclose migrant care workers in their journey and in their workplaces and ensure their return home after a placement. These new forms of migrant work affect not only care workers that are placed by home care agencies but also care work more generally. The emergence of the live-in care industry not only shapes the working conditions of those carers placed by home care agencies but also contributes to a formalisation of such working conditions and a changing understanding of home care work in society more generally. Thus, finding care work through private networks or through official home care agencies should not be seen as two separate spheres but as a continuum. Care agencies contribute to the production of an unequal division of care labour in the transnational context. This raises a further question: What do these developments mean for migration control?

Control of live-in care labour migration

It must have been a puzzling picture for the border guard: a woman in her mid-thirties, a young Asian-looking woman, and two older women in the back of a private vehicle. We handed him our Swiss and Slovakian identity cards. 'Aaah, car-pooling (*Mitfahrgemeinschaft*)?' he said upon looking at us. My fellow travellers nodded, and without further ado we rolled through the gate with a simple wave. It was 5 a.m. and we had been riding for around 11 hours in the car from southern Slovakia to Switzerland. The border guard most probably could not have known our reasons for travelling and the background of our constellation. His task was to check whether we had permission to travel to Switzerland and whether we had any goods to declare for customs. Since our passports were covered by the agreement on the free movement of workers, we had no difficulties in entering Switzerland.

The opening up of borders between the EU and EFTA states and the relaxation of admission regulations in Switzerland mark a crucial turn in migration policies and in migration control. This context of deterritorialised state borders raises the question of how migration control occurs. How has migration control become embedded in everyday life? At the beginning of this book, I argued that the migration infrastructure of live-in care work is crucial to the shifting power relations in the control of migration. In the following sections, I discuss the implications of the emergence of home care agencies for migration control and the consequences that may arise from these new developments.

Border control and differential inclusion of workers

While the freedom of movement between Switzerland and the EU-8 countries was only introduced in 2011, care workers had been able to travel to Germany and Austria to work legally as live-in care workers since 2004, when the accession states became members of the European Union (Krawietz 2014, 40–41). Presumably, care workers from Eastern Europe were able to travel to Switzerland through Germany and Austria without necessarily being checked at the borders, since Switzerland has gradually abandoned systematic identity checks at its borders since 2006. However, if care workers had been checked at the border as my car-pooling group was checked, their entry to Switzerland would have been denied. Little is known about the way that live-in care workers travelled to Switzerland before 2011 and how they found work in informal employment. What is clear is that their migration to Switzerland preceded state regulation (see Schilliger 2009). Live-in care work already existed through illegal entry and residency and irregular employment. Migration has its own logic, and as Andrijasevic (2009, 399–400) notes, border control and refusal of entry 'do not necessarily prevent or stop migratory movements', but simply 'decelerate the speed of migratory flows by momentarily diverting their directionality and regulating the time of migration'. Hence, it was just a question of time before the nationals of the accession states gained legal access to employment in the European Union and in Switzerland, as their freedom of movement 'has been delayed for a period of between two and seven years after accession' (Andrijasevic 2009, 300). In this sense, the relaxation in admission regulations with the enlargement of the zone of free movement to the EU-8 states can be understood as the timed differential inclusion of new workforce and is less about blocking migration than managing it.

'Shared sovereignty' in migration management

With the new liberalised admission regulations, it seems at first sight that everyone within the zone of free movement has equal access to the labour markets of the wealthier countries in the European Union. Furthermore, whether the free movement of labour and the Schengen system has moved us closer to a supranational and/or European citizenship and has rendered state sovereignty obsolete has been debated extensively (see Bauböck 2007, 2011; Dobson 2006; Keating 2001). The everyday practices, particularly in recruitment, that facilitate migrant care workers' journeys from their own homes to Switzerland show how the control of labour migration takes place not only at state borders but also at other scales and locations. The mass recruitment of labour in the post-war period was controlled through physical borders at the national boundaries, but these have now partly shifted away to new borders that are less visible. In addition, control of migration has partially been delegated from the state authorities to the recruitment agencies, home care agencies, and care recipients and their family members who select, place, and employ care workers and who organise their work schedules.

By engaging in selective recruitment of labour from Eastern Europe, these care agencies discriminate against prospective migrants from places other than there. The projected image of live-in care workers as mature, warm-hearted women from Eastern Europe has far-reaching consequences. It discriminates against men and younger women seeking access to the Swiss live-in care labour market. Furthermore, it perpetuates the stereotype of care work as the exclusive domain of women. Care agencies, an essential part of the migration infrastructure, actively contribute to the construction of a gendered migration channel from the EU-8 countries into the live-in care labour market in Switzerland through their recruitment and placement practices.

Although the passport as a marker of citizenship (Hindess 2000, 1487) no longer plays an immediate role in entry, the opening of the borders to EU-8 countries does not provide equal migration opportunities. Access to labour markets and livelihoods in Switzerland are still connected to political, social, and economic history. For example, care agencies can exhibit an implicit racism in their recruitment process, for instance in the exclusion of applicants of non-white physical appearance, which can only be understood in relation to Western Europe's colonial past (see Gutiérrez Rodriguez 2010; Rigo 2005), and the dual migration policy that excludes the admission of third-state nationals. Moreover, as I argue in Chapter 6, the recruitment of care workers incorporates existing discourses that assume essential differences between Western and Eastern Europe (see Kuus 2004; Melegh 2006).

The decentralisation and proliferation of borders begs the question of whether Switzerland as a nation state has lost its decision-making power over the admission of migrants. I argue that care agencies did not emerge as a bridge between an existing labour demand and a supply of care workers independently of the state, but they are to a certain extent institutionally shaped as gatekeepers. The state has instruments to influence the placement of live-in care workers and the gendered migration channel from Eastern Europe to Switzerland; for example, it regulates the care agencies and the working conditions of live-in care workers. It could, for instance, apply labour law to the working relations in households and hence create working conditions that are as attractive for local care workers as they are for migrant care workers. It could impose rigorous penalties on households that work with agencies from abroad, that employ care workers in irregular employment, and that fail to comply with the working conditions stipulated in standard employment contracts. The reasons for the state not doing so can only be explained by an interest in supporting the circular migration of live-in care workers from low-wage countries, and this may well be designed to avoid or delay discussion of the responsibilities of the state in elderly care. The desirability of migrant live-in care workers as a workforce to support Swiss households can be compared with the tolerance of irregularly employed *sans-papiers*, as Wicker (2010, 236–44) argues, as 'cheap labor ideally suited to the supply and demand structure of a flexible employment sector' in Switzerland.

In other words, the state has not lost sovereign power as legal authority over human movement. Live-in care workers are very much subject to the state's differentiated inclusion of migrants. Live-in care workers and home care agencies are obliged to register short-term or long-term residence permits with the state's central register of foreigners. The difference between this new approach and a more classical notion of state and territorial sovereignty lies in the 'growing involvement of non-state actors' to govern migration (Andrijasevic 2009, 396). The 'progressive deterritorialization of the internal and external borders of the European polity', as I have demonstrated by tracing the everyday practices that facilitate migrant care workers' journeys to Switzerland, 'allows sovereignty to be shared by different public and private actors' (Rigo 2005, 4).

Fundamental changes in migration control

In sum, my findings point to a fundamental change in migration in recent decades with crucial consequences for migrant workers. Whereas traditional attempts to control migration were determined by state governments and economic agreements and, even though this was not foreseen by the state, eventually resulted in permanent residence for migrants after gaining social rights, new migration patterns are now more strongly characterised by international policies, transnationally operated services, and the 'temporary mobility' of workers within the European Union and EFTA states (Bauloz 2016). This shift has been called the migration-mobility nexus, a convergence of the usage of the two terms denoting migration (see Bauloz 2016; and see NCCR 2017 for an overview of the projects on the topic).

Whereas migration to Switzerland in the past was intended to be of temporary nature, regulated by the state with restricted social rights, it is now characterised by freedom of movement and full social rights within the EU and EFTA states. Care workers from Eastern European countries are legally entitled to migrate to Switzerland with their families and to claim unemployment benefits. However, the circumstances of short-term stints, circular migration, isolated and difficult working conditions, and low salaries restrict their realistic opportunities to claim these social rights and to bring their families. The migration infrastructure and private for-profit home care agencies play a significant role in the (re)production of these circumstances.

To conclude, the findings indicate fundamental changes in migration control. Whereas labour migration control largely took place at physical borders before the agreement on the free movement of workers, it has now shifted to less visible borders around and within labour markets, such as the live-in care market. And it has partly been delegated from state authority to private actors such as home care agencies, recruitment agencies and care recipients, who now act as gatekeepers for care workers' access to live-in care work. The migration liberalisation within the EU and EFTA states has not only granted workers more rights to move freely for work but has also provided private actors such as home care agencies with new powers in determining who and under which conditions workers migrate. This does not mean that the state has become powerless as the legal authority over movement. However, state control has become more indirect, since it now regulates the agencies and their business practices and the working conditions of migrant live-in care workers. Hence, home care agencies play a key part in shaping migration patterns and the migration geographies of migrant live-in care workers in Europe.

Time for a care revolution

What does migration for live-in care work mean for society? A view I have frequently encountered both during fieldwork and in my everyday life relates to the argument that live-in care work performed by migrant workers is a win–win situation for all participants. If argued from an individual point of view, and if we only focus on live-in care work as opportunities for individuals to earn money and improve their personal economic and social situations, live-in care work can indeed be beneficial for these individuals. However, through the lens of the concept of migration infrastructure, I showed that the rise of placements of migrant live-in care workers is about much more.

The emergence of home care agencies and the growth of live-in care labour migration is embedded in various transformation processes: In the regulatory dimension, a liberalisation of migration regulation has gone hand in hand with the political promotion of circular migration within the European Union. Simultaneously, neoliberal practices led to a shift from public to private responsibility for care. In the technological dimension, developments in communication and information technology have proven highly relevant for the recruitment and marketing of live-in care for home care agencies and in the building and maintenance of new business and private relations. In the social dimension, various processes have merged: Firstly, a feminisation of the general workforce has contributed to a relative decrease of capacity for care in Switzerland. This development is conditioned by a gendered division of labour that attributes domestic and care work chiefly to women. Secondly, histories of migration in the EU-8 accession states have shaped local communities. Correspondingly, going abroad for live-in care work can be a common phenomenon in some places, and care workers share and circulate knowledge on the topic. In the dimension that Xiang and Lindquist denote as 'humanitarian' (Xiang & Lindquist 2014, 124), there has been an intensification of media attention on home care agencies and migrant care workers. This, as Schwiter, Pelzelmayer, and Thurnherr (2018) argue, produces a self-fulfilling prophecy and contributes to the diffusion of live-in care.

These interconnected developments in the social, regulatory, technological, and humanitarian dimensions have brought to the fore new commercial infrastructure and the development of home care agencies and other businesses, such as transport and recruitment companies. This in turn has required new regulations. Consequently, the government, as well as bodies such as trade unions and migrant workers' organisations, engage in direct struggles to improve the working and living conditions of live-in care workers. The emergence of home care agencies placing migrant live-in care workers and offering packaged home care services has triggered a range of new developments and discussions in the social and regulatory fields, focusing on how and who should care for the elderly in Switzerland in the future. Therefore, the rise of placement agencies and the intensification of live-in care placements are about significantly more than individual chances for migrant workers. Live-in care work in Switzerland involves current social transformations and a range of struggles that lie at the heart of these changes: the struggles of care workers to access work and to migrate autonomously, struggles around gender equality, and struggles around the value of care work. These struggles are all interconnected. The forces and infrastructures that enable migration are linked to societal negotiations of how to understand new family relations and responsibilities in domestic and care work.

Live-in care of the elderly by women migrant care workers in Switzerland is a model that is based on inequality in two main ways: an unequal distribution of care work between men and women and unequal working and living conditions between migrant workers and locals. It is a short-term solution that reinforces gender and socio-economic inequalities. A solution for society as a whole is only sustainable if it addresses these inequalities in the long run rather than merely the 'care-crisis' (Schilliger 2014, 99). This means that care work has to become more recognised. It also means that working conditions and salaries have to be improved so that workers caring for the elderly are able to secure an income that allows them to live in Switzerland. To be more precise, care work that is not based on such inequalities should enable care workers to work shifts and working hours comparable to those of local care workers in the health sector, to live out and to enable time and capacity for care workers to care for themselves and their own family members.

These issues concerning the value and responsibilities of domestic and care work involve not only care workers but society as a whole. With this in mind, several voices have called for 'an ethics of care' (McDowell 2009, 156–57) and, more recently, for a care revolution (Winker 2009, 2015). McDowell (2009, 157), for example, argues that the notion of an ethic of care as 'mutual obligations and relations of trust' should be applied not only to 'social relations in the familial and domestic arena' but also to 'the public sphere of the labour market'. Correspondingly, she calls for more solidarity in society and measures to create gender equality as well as 'a wider distribution of the responsibility for the labour of caring' (McDowell 2009, 157). In a similar way, the notion of care revolution addresses a need for fundamental social changes in relation to care. For this purpose, an action conference took place on the topic in Berlin in March 2014. Following the conference, a broad range of activists, groups, organisations, and researchers formed the network Care Revolution in Germany (Winker 2015). The supporters of the concept of a care revolution call for time sovereignty and existential security, an expansion of social infrastructure, and democratisation and self-governance in the care domain. These changes are important conditions for the development of both a caring society and an economy that addresses and satisfies the human needs of all people without placing a burden on people in other places of the world (Winker 2015).

Labour migration based on social and economic inequalities is never just a win–win situation. It is always intrinsically linked with migrant workers' capacities to be mobile and freedom of movement. It is embedded in social and socio-material relations and can only be understood in connection to wider social transformations. Hence, quests to explain and understand migration must always take into account how migration is enabled, by whom, and for what purpose: in short, how it is brokered.

Bibliography

Abrantes, Manuel. 2014. '"I Know It Sounds Nasty and Stereotyped": Searching for the Competent Domestic Worker: Searching for the Competent Domestic Worker'. *Gender, Work & Organization* 21 (5): 1–16. https://doi.org/10.1111/gwao.12046.

Adey, Peter. 2006. 'If Mobility Is Everything Then It Is Nothing: Towards a Relational Politics of (Im)Mobilities'. *Mobilities* 1 (1): 75–94. https://doi.org/10.1080/17450100500489080.

Akalin, Ayse. 2015. 'Motherhood as the Value of Labour: The Migrant Domestic Workers' Market in Turkey'. *Australian Feminist Studies* 30 (83): 65–81. https://doi.org/10.1080/08164649.2014.998451.

Allen, John, and Chris Hammnet, eds. 1999. *A Shrinking World? Global Unevenness and Inequality*. Reprinted as a paperback 1999. The Shape of the World. Oxford: Oxford University Press.

Alleva, Vania, and Pierre-Alain Niklaus. 2004. *Leben und arbeiten im Schatten. Die erste detaillierte Umfrage zu den Lebens-und Arbeitsbedingungen von Sans-Papiers in der Deutschschweiz*. Basel: Anlaufstelle für Sans-Papiers & Gewerkschaft Bau & Industrie GBI. www.sans-papiers.ch/fileadmin/redaktion/Hintergrund/1Studie_Arbeitsbedingungen_2004.pdf.

Ally, Shireen. 2005. 'Caring About Care Workers: Organizing in the Female Shadow of Globalization'. *Labour, Capital and Society/Travail, Capital et Société* 38 (1/2): 184–207. www.jstor.org/stable/43158592.

Amin, Ash. 2014. 'Lively Infrastructure'. *Theory, Culture & Society* 31 (7–8): 137–61. https://doi.org/10.1177/0263276414548490.

ANAG, Bundesgesetz über Aufenthalt und Niederlassung der Ausländer. 1931. www.admin.ch/opc/de/classified-compilation/19310017/200501010000/142.20.pdf.

Anderson, Bridget. 2000. *Doing the Dirty Work? The Global Politics of Domestic Labour*. London: Zed Books.

———. 2006. *Doing the Dirty Work? Migrantinnen Und Die Globalisierung Der Hausarbeit*. Berlin: Assoziation A.

Anderson, Bridget, and Isabel Shutes. 2014. *Migration and Care Labour. Theory, Policy and Politics*. Palgrave Macmillan. www.palgraveconnect.com/doifinder/10.1057/9781137319708.

Andrijasevic, Rutvica. 2009. 'Sex on the Move: Gender, Subjectivity and Differential Inclusion'. *Subjectivity* 29 (1): 389–406. https://doi.org/10.1057/sub.2009.27.

Aoyama, Yuko, James T. Murphy, and Susan Hanson. 2011. *Key Concepts in Economic Geography*. Key Concepts in Human Geography. Los Angeles, CA; London: SAGE.

AOZ. 2013a. *Care Migrantinnen in Schweizer Privathaushalten. Vor-Ort-Tour: Einblicke in Einen Wachsenden Prekären Arbeitsmarkt. Veranstaltung Des AOZs: Viel Markt – (Noch) Wenig Regulierung. Besuch Bei Einer Privaten Personalverleihagentur. Veranstaltung Der AOZ*. Zürich.

———. 2013b. *Care Migrantinnen in Schweizer Privathaushalten. Vor-Ort-Tour: Einblicke in Einen Wachsenden Prekären Arbeitsmarkt. Veranstaltung Des AOZs: Viel Markt – (Noch) Wenig Regulierung. Im Einsatz Für Bessere Arbeitsbedingungen. Erfahrungsberichte Einer Care Migrantin Und Gewerkschaftliche Positionen. Veranstaltun Der AOZ.* Zürich.

Arlettaz, Silvia. 2012. 'Saisonniers'. In *Historisches Lexikon der Schweiz*. www.hls-dhs-dss.ch/textes/d/D25738.php.

AVG. 1991. *SR 823.113 Verordnung vom 16. Januar 1991 über Gebühren, Provisionen und Kautionen im Bereich des Arbeitsvermittlungsgesetzes (Gebührenverordnung AVG, GebV-AVG)*. www.admin.ch/opc/de/classified-compilation/19910006/index.html.

AVV (Vorname). 1989. *SR 823.11 Bundesgesetz vom 6. Oktober 1989 über die Arbeitsvermittlung und den Personalverleih (Arbeitsvermittlungsgesetz, AVG). SR*. www.admin.ch/opc/de/print.html.

Ayalon, Liat. 2009. 'Beliefs and Practices Regarding Alzheimer's Disease and Related Dementias Among Filipino Home Care Workers in Israel'. *Aging & Mental Health* 13 (3): 456–62. https://doi.org/10.1080/13607860802534625.

———. 2010. 'The Perspectives of Older Care Recipients, Their Family Members, and Their Round-the-Clock Foreign Home Care Workers Regarding Elder Mistreatment'. *Aging & Mental Health* 14 (4): 411–15. https://doi.org/10.1080/13607860903586110.

Ayalon, Liat, Miri Kaniel, and Liat Rosenberg. 2008. 'Social Workers' Perspectives on Care Arrangements Between Vulnerable Elders and Foreign Home Care Workers: Lessons from Israeli/Filipino Caregiving Arrangements'. *Home Health Care Services Quarterly* 27 (2): 121–42. https://doi.org/10.1080/01621420802022563.

Bachinger, Almut. 2009. 'Der irreguläre Pflegearbeitsmarkt. Zum Transformationsprozess von unbezahlter in bezahlte Arbeit durch die 24-Stunden-Pflege'. Dissertation, Wien: Universität Wien. http://othes.univie.ac.at/8038/.

———. 2010. '24-Stunden-Betreuung-gelungenes Legalisierungsprojekt oder prekäre Arbeitsmarktintegration?' *SWS-Rundschau* 50 (4): 399–412. www.ssoar.info/ssoar/handle/document/33968.

Bahna, Miloslav, and Martina Sekulová. 2019. *Crossborder Care: Lessons from Central Europe*. Cham: Springer International Publishing. https://doi.org/10.1007/978-3-319-97028-8.

Bakan, Abigail B., and Daiva K. Stasiulis. 1995. 'Making the Match: Domestic Placement Agencies and the Racialization of Women's Household Work'. *Signs* 20 (2): 303–35. www.jstor.org/stable/3174951.

Baldassar, Loretta, and Laura Merla, eds. 2014. *Transnational Families, Migration and the Circulation of Care: Understanding Mobility and Absence in Family Life*. Routledge Research in Transnationalism 29. New York, NY: Routledge.

Batthyany, Béla. 2013. '"Hilfe aus dem Osten" – Pflegemigrantinnen in der Schweiz'. *Dokumentarfilm des Schweizer Fernsehens SRF. Erstausstrahlung am 20.06.2013 auf SRF1.*

Bauböck, Rainer. 2007. 'Why European Citizenship? Normative Approaches to Supranational Union'. *Theoretical Inquiries in Law* 8 (2): 453–88. www.degruyter.com/view/j/til.2007.8.issue-2/til.2007.8.2.1157/til.2007.8.2.1157.xml.

———. 2011. 'Temporary Migrants, Partial Citizenship and Hypermigration'. *Critical Review of International Social and Political Philosophy* 14 (5): 665–93. https://doi.org/10.1080/13698230.2011.617127.

Bauloz, Céline. 2016. 'Blurred Lines: Migration and Mobility in EU Law and Policy'. Working Paper 4. NCCR on the Move. Neuchâtel: University of Neuchatel. http://nccr-onthemove.ch/wp_live14/wp-content/uploads/2013/01/nccrotm-WP4-Bauloz-Blurred-Lines.pdf.

Benner, Chris. 2003. 'Labour Flexibility and Regional Development: The Role of Labour Market Intermediaries'. *Regional Studies* 37 (6–7): 621–33. https://doi.org/10.1080/00 34340032000108723.

Benner, Chris, Laura Leete, and Manuel Pastor. 2007. *Staircases or Treadmills? Labor Market Intermediaries and Economic Opportunity in a Changing Economy.* Russell Sage Foundation. New York.

BFS, Bundesamt für Statistik. 2017. *Löhne, Erwerbseinkommen und Arbeitskosten.* www.bfs.admin.ch/bfs/de/home/statistiken/arbeit-erwerb/loehne-erwerbseinkommen-arbeitskosten.html.

BGE 124 III 249 E.3. n.d. Accessed 30 May 2017. www.servat.unibe.ch/Dfr/bge/c3124249. html.

Bock, Gisela, and Barbara Duden. 1976. 'Arbeit aus Liebe – Liebe als Arbeit. Zur Entstehung der Hausarbeit im Kapitalismus'. *Frauen und Wissenschaft. Beiträge zur Berliner Sommeruniversität für Frauen* 2: 118–99.

Boris, Eileen, and Jennifer N. Fish. 2014. ' "Slaves No More": Making Global Labor Standards for Domestic Workers'. *Feminist Studies* 40 (2): 411–43. www.jstor.org/stable/10.15767/feministstudies.40.2.411.

Bracher, Katharina. 2011. 'Eine Altenpflegerin für weniger als 2000 Franken'. *NZZ*, 13 March 2011.

B,S,S., Volkswirtschaftliche Beratung. 2016. *24-Stunden-Betagtenbetreuung in Privathaushalten. Regulierungsfolgenabschätzung zu den Auswirkungen der Lösungswege gemäss Bericht zum Postulat Schmid-Federer 12.3266. Pendelmigration zur Alterspflege. Schlussbericht.* Basel.

Bundesrat. 2011. *Bundesratsbeschluss über die Allgemeinverbindlicherklärung des Gesamtarbeitsvertrages für den Personalverleih vom 13. Dezember 2011.* www.seco.admin.ch/seco/de/home/Arbeit/Personenfreizugigkeit_Arbeitsbeziehungen/Gesamtarbeitsvertraege_Normalarbeitsvertraege/Gesamtarbeitsvertraege_Bund/Allgemeinverbindlich_erklaerte_Gesamtarbeitsvertraege/Personalverleih.html.

———. 2013. 'Erläuternder Bericht zum Entwurf für die Verlängerung und Änderung des Normalarbeitsvertrages für Arbeitnehmerinnen und Arbeitnehmer in der Hauswirtschaft'. Bern. www.news.admin.ch/NSBSubscriber/message/attachments/32680.pdf.

———. 2017a. '24-Stunden-Betreuungsarbeit: Neue Regelung bis Mitte 2018'. www. admin.ch/gov/de/start/dokumentation/medienmitteilungen.msg-id-67221.html.

———. 2017b. 'Erläuternder Bericht zum Entwurf für die Verlängerung und Änderung des Normalarbeitsvertrages für Arbeitnehmerinnen und Arbeitnehmer in der Hauswirtschaft'. Bern. www.newsd.admin.ch/newsd/message/attachments/46596.pdf.

———. 2018. 'Bundesrat Will Personenfreizügigkeit Nicht Kündigen'. www.admin.ch/gov/de/start/dokumentation/medienmitteilungen.msg-id-73208.html.

———. 2019. 'Umsetzung der Initiative "Gegen Masseneinwanderung" '. www.admin. ch/gov/en/start/documentation/dossiers/umsetzung-der-initiative-gegen-masseneinwanderung.html?_organization=1&_topic=102&_startDate=01.01.2014&_pageIndex=0.

Büscher, Monika, and John Urry. 2009. 'Mobile Methods and the Empirical'. *European Journal of Social Theory* 12 (1): 99–116. https://doi.org/10.1177/1368431008099642.

Büscher, Monika, John Urry, and Katian Witchger, eds. 2011. *Mobile Methods.* Abingdon, Oxon; New York, NY: Routledge.

CareInfo. 2013. 'CareInfo'. http://care-info.ch/de/.

Castles, Stephen. 2006. 'Guestworkers in Europe: A Resurrection?' *International Migration Review* 40 (4): 741–66. https://doi.org/10.1111/j.1747-7379.2006.00042.x.

Charmaz, Kathy. 2014. *Constructing Grounded Theory*. 2nd ed. Introducing Qualitative Methods. London; Thousand Oaks, CA: Sage.

Chau, Huey Shy. 2019. 'Producing (Im)Mobilities in Home Care for the Elderly: The Role of Home Care Agencies in Switzerland'. *International Journal of Ageing and Later Life* (August): 1–28. https://doi.org/10.3384/ijal.1652-8670.18396.

Chau, Huey Shy, Awanish Kumar, and Silva Lieberherr. 2015. 'Non-Family Labour in the Swiss Agriculture: A Status Report and Future Prospects'. *Journal of Socio-Economics in Agriculture* 7 (1): 1–10. www.sse-sga.ch/_downloads/YSA2014_Chau.pdf.

———. 2018. 'Short-Term Circular Migration and Gendered Negotiation of the Right to the City: The Case of Migrant Live-in Care Workers in Basel, Switzerland'. *Cities* 76 (June): 4–11. https://doi.org/10.1016/j.cities.2017.04.004.

Chen, Martha Alter. 2011. 'Recognizing Domestic Workers, Regulating Domestic Work: Conceptual, Measurement, and Regulatory Challenges'. *Canadian Journal of Women and the Law/Revue Femmes et Droit* 23 (1): 167–84. http://utpjournals.metapress.com/index/U80107514G596568.pdf.

Cheng, Shu-Ju Ada. 1996. 'Migrant Women Domestic Workers in Hong Kong, Singapore and Taiwan: A Comparative Analysis'. *Asian and Pacific Migration Journal* 5 (1): 139–52. http://journals.sagepub.com/doi/abs/10.1177/011719689600500107.

Christensen, Karen, and Doria Pilling, eds. 2018. *The Routledge Handbook to Social Care Work around the World*. Abingdon, Oxon; New York, NY: Routledge.

Coe, Neil M., Jennifer Johns, and Kevin Ward. 2009. 'Agents of Casualization? The Temporary Staffing Industry and Labour Market Restructuring in Australia'. *Journal of Economic Geography* 9 (1): 55–84. https://doi.org/10.1093/jeg/lbn029.

———. 2012. 'Managed Flexibility: Labour Regulation, Corporate Strategies and Market Dynamics in the Swedish Temporary Staffing Industry'. *European Urban and Regional Studies* 16 (1) (December): 65–85.

Constable, Nicole. 1997a. *Maid to Order in Hong Kong: Stories of Filipina Workers*. Ithaca, NY: Cornell University Press.

———. 1997b. 'Sexuality and Discipline Among Filipina Domestic Workers in Hong Kong'. *American Ethnologist* 24 (3): 539–58. http://onlinelibrary.wiley.com/doi/10.1525/ae.1997.24.3.539/full.

———. 2009. 'The Commodification of Intimacy: Marriage, Sex, and Reproductive Labor'. *Annual Review of Anthropology* 38 (1): 49–64. https://doi.org/10.1146/annurev.anthro.37.081407.085133.

Cox, Nicole, and Silvia Federici. 1976. *Counter-Planning from the Kitchen: Wages for Housework, a Perspective on Capital and the Left*. Published jointly by New York Wages for Housework Committee and Falling Wall Press.

Cox, Rosie. 2006. *The Servant Problem: Domestic Employment in a Global Economy*. London; New York, NY: I.B. Tauris.

Crane, Randall. 2007. 'Is There a Quiet Revolution in Women's Travel? Revisiting the Gender Gap in Commuting'. *Journal of the American Planning Association* 73 (3): 298–316.

Cresswell, Tim. 2006. *On the Move: Mobility in the Modern Western World*. New York, NY: Routledge.

———. 2010. 'Towards a Politics of Mobility'. *Environment and Planning D: Society and Space* 28 (1): 17–31. https://doi.org/10.1068/d11407.

Cresswell, Tim, Sara Dorow, and Sharon Roseman. 2016. 'Putting Mobility Theory to Work: Conceptualizing Employment-Related Geographical Mobility'. *Environment*

    and Planning A: Economy and Space 48 (9): 1787–803. https://doi.org/10.1177/0308
518X16649184.

Cuban, Sondra. 2013. *Deskilling Migrant Women in the Global Care Industry*. London:
Palgrave Macmillan UK. https://doi.org/10.1057/9781137305619.

Cuban, Sondra, and Corinne Fowler. 2012. 'Carers Cruising Cumbria and Meals on the
Mile: The Drive of Migrants in Fieldwork and Fiction'. *Mobilities* 7 (2): 295–315.
https://doi.org/10.1080/17450101.2012.654998.

D'Amato, Gianni. 2008. 'Erwünscht, aber nicht immer willkommen. Die Geschichte der
Einwanderungspolitik'. In *Die neue Zuwanderung: die Schweiz zwischen Brain-gain
und Überfremdungsangst*, edited by Daniel Müller-Jentsch. Zürich: Avenir Suisse.

De Coss-Corzo, Julio Alejandro. 2016. 'Thinking Infrastructure as a Contested Political
Space: Theoretical Reflections and Methodological Implications'. Presented at the BISA
Conference, Edinburgh, Scotland.

Denzin, Norman K., and Yvonna S. Lincoln, eds. 2011. *The Sage Handbook of Qualitative
Research*. 4th ed. Thousand Oaks, California: Sage.

Díaz-Pinés, Agustín. 2009. 'Indicators of Broadband Coverage'. OECD Digital Economy Papers. Paris: Organisation for Economic Co-operation and Development. www.
oecd-ilibrary.org/content/workingpaper/5kml8rfg777l-en.

Dobson, Lynn. 2006. *Supranational Citizenship*. Manchester; New York, NY: Manchester
University Press.

Donath, Susan. 2000. *The Other Economy: A Suggestion for a Distinctively Feminist Economics*. *Feminist Economics* 6 (1): 115–123. https://doi.org/10.1080/135457000337723.

Dorow, Sara, and Shingirai Mandizadza. 2018. 'Gendered Circuits of Care in the Mobility
Regime of Alberta's Oil Sands'. *Gender, Place & Culture* 25 (8): 1–16. https://doi.org/1
0.1080/0966369X.2018.1425287.

Dorow, Sara, Sharon R. Roseman, and Tim Cresswell. 2017. 'Re-Working Mobilities:
Emergent Geographies of Employment-Related Mobility'. *Geography Compass* 11
(12): 1–12. https://doi.org/10.1111/gec3.12350.

Eelens, F., and J.D. Speckmann. 1990. 'Recruitment of Labor Migrants for the Middle
East: The Sri Lankan Case'. *The International Migration Review* 24: 297–322.

EGB, Eidgenössisches Büro für Gleichstellung von Frau und Mann. 2010. *Anerkennung
und Aufwertung der Care-Arbeit. Impulse aus Sicht der Gleichstellung*. Bern: Eidgenössisches Büro für Gleichstellung von Frau und Mann.

Ehrenreich, Barbara, and Arlie Hochschild. 2003. *Global Woman: Nannies, Maids and Sex
Workers in the New Economy*. London: Granta Books.

EKM, Eidgenössische Migrationskommission. 2015. 'Initiativen zur Begrenzung
der Zuwanderung und gegen Überfremdung'. www.ekm.admin.ch/ekm/de/home/
zuwanderung – aufenthalt/zuwanderung/geschichtliches/volksinitiativen.html.

Elias, Juanita. 2008. 'Struggles over the Rights of Foreign Domestic Workers in Malaysia:
The Possibilities and Limitations of "Rights Talk"'. *Economy and Society* 37 (2): 282–
303. https://doi.org/10.1080/03085140801933330.

Ellner, Susanna. 2011. '"Pendelmigrantinnen" als Hausangestellte'. *NZZ*, 12 November
2011.

Elrick, Tim. 2008. 'The Influence of Migration on Origin Communities: Insights from
Polish Migrations to the West'. *Europe-Asia Studies* 60 (9): 1503–17. https://doi.
org/10.1080/09668130802362243.

———. 2009. 'Transnational Networks of Eastern European Labour Migrants'. PhD, Berlin: Freie Universität Berlin. www.diss.fu-berlin.de/diss/servlets/MCRFileNodeServlet/
FUDISS_derivate_000000005784/Elrick_Doktorarbeit.pdf.

Elrick, Tim, and Emilia Lewandowska. 2008. 'Matching and Making Labour Demand and Supply: Agents in Polish Migrant Networks of Domestic Elderly Care in Germany and Italy'. *Journal of Ethnic and Migration Studies* 34 (5): 717–34.

England, Kim, and Isabel Dyck. 2012. 'Migrant Workers in Home Care: Routes, Responsibilities, and Respect'. *Annals of the Association of American Geographers* 102 (5): 1076–83. https://doi.org/10.1080/00045608.2012.659935.

EntsG, Entsendegesetz. 1999. *SR 823.20 Bundesgesetz vom 8. Oktober 1999 über die flankierenden Massnahmen bei entsandten Arbeitnehmerinnen und Arbeitnehmern und über die Kontrolle der in Normalarbeitsverträgen vorgesehenen Mindestlöhne. 823.20.* www.admin.ch/opc/de/classified-compilation/19994599/index.html.

EntsV. 2003. *SR 823.201 Verordnung vom 21. Mai 2003 über die in die Schweiz entsandten Arbeitnehmerinnen und Arbeitnehmer (EntsV). 823.201.* www.admin.ch/opc/de/classified-compilation/20030526/index.html.

European Commission. 2015. 'Posted Workers – Employment, Social Affairs & Inclusion – European Commission'. http://ec.europa.eu/social/main.jsp?catId=471.

European Migration Network. 2011. *Temporary and Circular Migration: Empirical Evidence, Current Policy Practice and Future Options in EU Member States.* Luxembourg: Publications Office of the European Union. https://ec.europa.eu/home-affairs/sites/homeaffairs/files/what-we-do/networks/european_migration_network/reports/docs/emn-studies/circular-migration/0a_emn_synthesis_report_temporary__circular_migration_final_sept_2011_en.pdf.

European Parliament. 2006. 'Legislative Resolution on the Proposal for a Directive of the European Parliament and of the Council on Services in the Internal Market'. Strasbourg. www.europarl.europa.eu/sides/getDoc.do?language=EN&reference=P6-TA-2006-0061&type=TA#top.

Eurostat. 2019. 'Living Conditions in Europe – Income Distribution and Income Inequality – Statistics Explained'. Eurostat. Statistics Explained. 2019. https://ec.europa.eu/eurostat/statistics-explained/index.php?title=Living_conditions_in_Europe_-_income_distribution_and_income_inequality.

Faist, Thomas. 2012. 'Toward a Transnational Methodology: Methods to Address Methodological Nationalism, Essentialism, and Positionality'. *Revue européenne des migrations internationales* 28 (1): 51–70. https://doi.org/10.4000/remi.5761.

FDF, Federal Department of Finance. 2016. *Langfristperspektiven der öffentlichen Finanzen in der Schweiz 2016.* Bern: Eidgenössisches Finanzdepartement. www.efd.admin.ch/efd/de/home/dokumentation/broschueren/periodika/langfristperspektiven-der-oeffentlichen-finanzen-in-der-schweiz-.html.

FDFA, Federal Department of Foreign Affairs. 2019. 'Free Movement of Persons – Functioning and Current State of Play'. Mission of Switzerland to the European Union. 2019. www.eda.admin.ch/missions/mission-eu-brussels/en/home/dossiers/personenfreizuegigkeit.html.

Fernandez, Bina, and Marina De Regt. 2014. *Migrant Domestic Workers in the Middle East: The Home and the World.* Basingstoke: Palgrave Macmillan.

Findlay, Allan, David McCollum, Sergei Shubin, Elina Apsite, and Zaiga Krisjane. 2012. 'The Role of Recruitment Agencies in Producing the "good" Migrant'. *Social & Cultural Geography* 14 (2): 145–67. https://doi.org/10.1080/14649365.2012.737008.

———. 2013. 'The Role of Recruitment Agencies in Producing the "good" Migrant'. *Social & Cultural Geography* 14 (2) (December): 145–67.

Fish, Jennifer N., and Eileen Boris. 2015. 'Decent Work for Domestics: Feminist Organizing, Worker Empowerment, and the ILO'. In *Towards a Global History of Domestic and Caregiving Workers*, edited by Dirk Hoerder, Elise Nederveen Meerkerk, and Silke Neunsinger, 530–52. Brill. Leiden, Boston.

Flückiger, Yves. 2008. *Le travail domestique en Suisse*. Geneva: Observatoire Universitaire de l'Emploi. Genève: Université de Genève. www.seco.admin.ch/seco/fr/home/Publikationen_Dienstleistungen/Publikationen_und_Formulare/Arbeit/Personen-freizuegigkeit_und_Arbeitsbeziehungen/Studien_und_Berichte/domestic-work-in-switzerland.html.

Flückiger, Yves, and Cyril Pasche. 2005. *Analyse du secteur clandestin de l'économie demestique à Genève*. Genève: Observatoire Universitaire de l'Emploi. www.sans-papiers.ch//fileadmin/redaktion/Hintergrund/1FRStudie_Hauswirtschaft_Genf_2004.pdf.

Folbre, Nancy. 1995. ' "Holding Hands at Midnight": The Paradox of Caring Labor'. *Feminist Economics* 1 (1): 73–92. https://doi.org/10.1080/714042215.

Fraser, Nancy. 2014a. 'Behind Marx's Hidden Abode'. *New Left Review* 86: 18.

———. 2014b. 'Can Society Be Commodities All the Way Down? Post-Polanyian Reflections on Capitalist Crisis'. *Economy and Society* 43 (4): 541–58. https://doi.org/10.1080/03085147.2014.898822.

———. 2016. 'Contradictions of Capital and Care'. *New Left Review* 100: 99–117.

Freedonia. 2017. 'Elder Care Services – Market Size, Market Share, Market Leaders, Demand Forecast, Sales, Company Profiles, Market Research, Industry Trends and Companies'. www.freedoniagroup.com/industry-study/elder-care-services-3214.htm.

Fürst, Christine. 2013. 'Ausländerinnen füllen die Lücke'. *Der Sonntag*, 19 February 2013.

Future Market Insights. 2019. 'Senior In-Home Care Services Market (2017–2027) Global Industry Analysis, Size and Forecast'. www.futuremarketinsights.com/reports/senior-in-home-care-services-market.

Gaetano, Arianne M., and Brenda S. A. Yeoh. 2010. 'Introduction to the Special Issue on Women and Migration in Globalizing Asia: Gendered Experiences, Agency, and Activism: Women and Migration in Globalizing Asia'. *International Migration* 48 (6): 1–12. https://doi.org/10.1111/j.1468-2435.2010.00648.x.

GAV. 2012. *Gesamtarbeitsvertrag Personalverleih. 21. Dezember 2011*. www.tempservice.ch/tempservice/mm/GAV_Personalverleih_13_12_2011_de.pdf.

GCIM. 2005. *Migration in an Interconnected World: New Directions for Action. Report of the Global Commission on International Migration*. Geneva: Global Commission on International Migration. www.unitar.org/ny/sites/unitar.org.ny/files/GCIM%20Report%20%20PDF%20of%20complete%20report.pdf.

Gertler, M. S. 2003. 'Tacit Knowledge and the Economic Geography of Context, or The Undefinable Tacitness of Being (There)'. *Journal of Economic Geography* 3 (1): 75–99. https://doi.org/10.1093/jeg/3.1.75.

Glenn, Evelyn Nakano. 1992. 'From Servitude to Service Work: Historical Continuities in the Racial Division of Paid Reproductive Labor'. *Signs*, 1–43. www.jstor.org/stable/3174725.

Global Market Insights. 2018. 'Geriatric Care Services Market Share Trends Growth Report 2018–2024'. www.gminsights.com/pressrelease/geriatric-care-services-market-size.

Goh, Charmian, Kellynn Wee, and Brenda S. A. Yeoh. 2016. 'Who's Holding the Bomb? Debt-Financed Migration in Singapore's Domestic Work Industry'. Working Paper 38. Migrating out of Poverty. Research Programme Consortium. University of Sussex.

———. 2017. 'Migration Governance and the Migration Industry in Asia: Moving Domestic Workers from Indonesia to Singapore'. *International Relations of the Asia-Pacific* 17 (3): 401–33. https://doi.org/10.1093/irap/lcx010.

Götz, Georg. 2013. 'Competition, Regulation, and Broadband Access to the Internet'. *Telecommunications Policy* 37 (11): 1095–109. https://doi.org/10.1016/j.telpol.2013.03.001.

Greuter, Susy. 2010. 'Langzeitpflege, Service public und der Druck der Ökonomisierung'. In *Zu gut für den Kapitalismus – Blockierte Potenziale in einer überforderten Wirtschaft: Denknetz Jahrbuch 2010*, edited by Denknetz, 106–12. Zürich: Edition 8.

Greuter, Susy, and Sarah Schilliger. 2010. '"Ein Engel aus Polen": Globalisierter Arbeitsmarkt im Privathaushalt von Pflegebedürftigen'. In *Krise. Lokal, global, fundamental: Denknetz Jahrbuch 2009*, edited by Denknetz. Zürich: Edition 8.

Guevarra, Anna Romina. 2010. *Marketing Dreams, Manufacturing Heroes: The Transnational Labor Brokering of Filipino Workers*. New Brunswick, NJ: Rutgers University Press.

Gutiérrez Rodriguez, Encarnación. 2010. *Migration, Domestic Work and Affect: A Decolonial Approach on Value and the Feminization of Labor*. Routledge Research in Gender and Society 26. New York, NY: Routledge.

Haidinger, Bettina. 2013. *Hausfrau für zwei Länder sein. Zur Reproduktion des transnationalen Haushalts*. Münster: Westfälisches Dampfboot.

Hanson, Susan, and Ibipo Johnston. 1985. 'Gender Differences in Work-Trip Length: Explanations and Implications'. *Urban Geography* 6 (3): 193–219. https://doi.org/10.2747/0272-3638.6.3.193.

Hanson, Susan, and Geraldine Pratt. 1995. *Gender, Work, and Space*. New York, NY: Psychology Press.

Harvey, David. 1989. *The Condition of Postmodernity: An Enquiry into the Origins of Social Change*. Malden, MA: Blackwell.

Harvey, Penelope, Casper Bruun Jensen, and Atsuro Morita, eds. 2017. 'Introduction: Infrastructural Complications'. In *Infrastructures and Social Complexity: A Companion*, 1–22. London; New York, NY: Routledge, Taylor & Francis Group.

Hess, Sabine. 2005. *Globalisierte Hausarbeit. Au-pair als Migrationsstrategie von Frauen aus Osteuropa*. Wiesbaden: Verlag für Sozialwissenschaften.

Hindess, Barry. 2000. 'Citizenship in the International Management of Populations'. *American Behavioral Scientist* 43 (9): 1486–97. https://doi.org/10.1177/00027640021956008.

Hochschild, Arlie R. 1979. 'Emotion Work, Feeling Rules, and Social Structure'. *American Journal of Sociology* 85 (3): 551–75. https://doi.org/10.2307/2778583.

———. 2000. 'Global Care Chains and Emotional Surplus Value'. In *On the Edge: Globalization and the New Millennium*, edited by Anthony Giddens and Will Hutton, 130–46. London: Sage Publishers.

———. 2003a. 'Love and Gold'. In *Global Women: Nannies, Maids and Sex Workers in the New Economy*, edited by Barbara Ehrenreich and Arlie R. Hochschild, 15–30. New York, NY: Metropolitan Books/Holt.

———. 2003b. *The Managed Heart: Commercialization of Human Feeling*. 20th anniversary ed. Berkeley, CA: University of California Press.

Hoffmann-Nowotny, Hans-Joachim. 1985. 'Switzerland'. In *European Immigration Policy: A Comparative Study*, edited by Tomas Hammar, 206–36. Cambridge: Cambridge University Press.

Hondagneu-Sotelo, Pierrette. 1997. *Domestic Employment Agencies in Los Angeles*. Los Angeles: University of Southern California, Southern California Studies Center.

———. 2000. *Doméstica: Immigrant Workers Cleaning and Caring in the Shadows of Affluence*. Berkeley: University of California Press.

Hondagneu-Sotelo, Pierrette, and Ernestine Avila. 1997. '"I'm Here, but I'm There": The Meanings of Latina Transnational Motherhood'. *Gender and Society* 11 (5): 548–71. www.jstor.org/stable/190339.

Höpflinger, Francois. 2011. *Pflegebedürftigkeit und Langzeitpflege im Alter: aktualisierte Szenarien für die Schweiz*. Bern: Huber.

Horváth, István. 2008. 'The Culture of Migration of Rural Romanian Youth'. *Journal of Ethnic and Migration Studies* 34 (5): 771–86. https://doi.org/10.1080/13691830802106036.

Huang, Shirlena, Leng Leng Thang, and Mika Toyota. 2012. 'Transnational Mobilities for Care: Rethinking the Dynamics of Care in Asia'. *Global Networks* 12 (2): 129–34. https://doi.org/10.1111/j.1471-0374.2012.00343.x.

Huang, Shirlena, and Brenda S. A. Yeoh. 1996. 'Ties That Bind: State Policy and Migrant Female Domestic Helpers in Singapore'. *Geoforum* 27 (4): 479–93. www.sciencedirect.com/science/article/pii/S0016718596000231.

———. 1998. 'Maids and Ma'ams in Singapore: Constructing Gender and Nationality in the Transnationalization of Paid Domestic Work'. *Geography Research Forum* 18: 22–48. http://raphael.geography.ad.bgu.ac.il/ojs/index.php/GRF/article/view/185.

———. 2007. 'Emotional Labour and Transnational Domestic Work: The Moving Geographies of "Maid Abuse" in Singapore'. *Mobilities* 2 (2): 195–217.

Huang, Shirlena, Brenda S. A. Yeoh, and Mika Toyota. 2012. 'Caring for the Elderly: The Embodied Labour of Migrant Care Workers in Singapore'. *Global Networks* 12 (2): 195–215. http://onlinelibrary.wiley.com/doi/10.1111/j.1471-0374.2012.00347.x/abstract.

IGA. 2007. *Sektoranalyse. Externe Haushaltsarbeit im Kanton Basel*. Basel: Interprofessionelle Gewerkschaft. www.sans-papiers.ch/fileadmin/redaktion/Hintergrund/1bSektoranalyse_externe_Hausarbeit_2007_05.pdf.

Irek, Małgorzata. 1998. *Der Schmugglerzug: Warschau – Berlin – Warschau; Materialien einer Feldforschung*. Berlin: Das Arabische Buch.

Karakayali, Juliane. 2010. 'Die Regeln Des Irregulären – Häusliche Pflege in Zeiten Der Globalisierung'. *Transnationale Sorgearbeit. Rechtliche Rahmenbedingungen Und Gesellschaftliche Praxis* (September): 151–69.

Kaufmann, Vincent. 2002. *Re-Thinking Mobility: Contemporary Sociology*. Aldershot, Hants: Ashgate.

Kaufmann, Vincent, Manfred Max Bergman, and Dominique Joye. 2004. 'Motility: Mobility as Capital'. *International Journal of Urban and Regional Research* 28 (4): 745–56. https://doi.org/10.1111/j.0309-1317.2004.00549.x.

Keating, Michael. 2001. *Plurinational Democracy: Stateless Nations in a Post-Sovereignty Era*. Oxford, England; New York, NY: Oxford University Press.

Kern, Alice, and Ulrike Müller-Böker. 2015. 'The Middle Space of Migration: A Case Study on Brokerage and Recruitment Agencies in Nepal'. *Geoforum* 65 (October): 158–69. https://doi.org/10.1016/j.geoforum.2015.07.024.

Killias, Olivia. 2009. 'The Politics of Bondage in the Recruitment, Training and Placement of Indonesian Migrant Domestic Workers'. *Sociologus* 59 (2): 145–72. http://ejournals.duncker-humblot.de/doi/abs/10.3790/soc.59.2.145.

Kislig, Bernhard. 2015. 'Illegale Anbieter locken mit günstigen Pflege-Angeboten'. *Berner Zeitung*, 31 March 2015.

Knobloch, Ulrike. 2009. 'Sorgeökonomie Als Allgemeine Wirtschaftstheorie'. *Olympe. Feministische Arbeitshefte Zur Politik* 30: 27–36.

Kofman, Eleonore, and Parvati Raghuram. 2015. *Gendered Migrations and Global Social Reproduction*. Migration, Minorities and Citizenship. Houndmills, Basingstoke, Hampshire; New York, NY: Palgrave Macmillan.

Krawietz, Johanna. 2010. 'Pflegearbeit unter Legitimationsdruck. Vermittlungsagenturen im transnationalen Organisationsfeld'. In *Transnationale Sorgearbeit. Rechtliche*

Rahmenbedingungen und gesellschaftliche Praxis, edited by Kirsten Scheiwe and Johanna Krawietz, 249–75. Wiesbaden: VS Verlag.

———. 2014. *Pflege grenzüberschreitend organisieren. Eine Studie zur transnationalen Vermittlung von Care-Arbeit*. Frankfurt am Main: Mabuse-Verlag.

Krissman, Fred. 2005. 'Sin Coyote Ni Patron: Why the "Migrant Network" Fails to Explain International Migration'. International Migration Review 39 (1): 4–44. https://doi.org/10.1111/j.1747-7379.2005.tb00254.x.

Kuus, Merje. 2004. 'Europe's Eastern Expansion and the Reinscription of Otherness in East-Central Europe'. *Progress in Human Geography* 28 (4): 472–89. https://doi.org/10.1 191/03091 32504ph498oa.

Lan, Pei-Chia. 2006. *Global Cinderellas: Migrant Domestics and Newly Rich Employers in Taiwan*. Durham; London: Duke University Press.

Larkin, Brian. 2013. 'The Politics and Poetics of Infrastructure'. *Annual Review of Anthropology* 42 (1): 327–43. https://doi.org/10.1146/annurev-anthro-092412-155522.

Liang, Li-Fang. 2011. 'The Making of an "Ideal" Live-in Migrant Care Worker: Recruiting, Training, Matching and Disciplining'. *Ethnic and Racial Studies* 34 (11): 1815–34. https://doi.org/10.1080/01419870.2011.554571.

Lin, Stephen, and Danièle Bélanger. 2012. 'Negotiating the Social Family: Migrant Live-in Elder Care-Workers in Taiwan'. *Asian Journal of Social Science* 40 (3): 295–320. https://doi.org/10.1163/156853112X650854.

Lin, Weiqiang, Johan Lindquist, Biao Xiang, and Brenda S. A. Yeoh. 2017. 'Migration Infrastructures and the Production of Migrant Mobilities'. *Mobilities* 12 (2): 1–8. https://doi.org/10.1080/17450101.2017.1292770.

Lindquist, Johan. 2010. 'Labour Recruitment, Circuits of Capital and Gendered Mobility: Reconceptualizing the Indonesian Migration Industry'. *Pacific Affairs* 83 (1): 115–32. www.ingentaconnect.com/content/paaf/paaf/2010/00000083/00000001/art00006.

———. 2012. 'The Elementary School Teacher, the Thug and His Grandmother: Informal Brokers and Transnational Migration from Indonesia'. *Pacific Affairs* 85 (1): 69–89. https://doi.org/10.5509/201285169.

Lindquist, Johan, and Biao Xiang. 2018. 'The Infrastructural Turn in Asian Migration'. In *Routledge Handbook of Asian Migrations*, edited by Gracia Liu-Farrer and Brenda S. A. Yeoh, 152–61. New York, NY: Routledge.

Lindquist, Johan, Biao Xiang, and Brenda S. A. Yeoh. 2012. 'Opening the Black Box of Migration: Brokers, the Organization of Transnational Mobility and the Changing Political Economy in Asia'. *Pacific Affairs* 85 (1): 7–19. https://doi.org/10.5509/20128517.

Lockrem, Jessica, and Adonia Lugo. 2012. 'Infrastructure: Editorial Introduction. Curated Collections'. *Cultural Anthropology* (26 November). https://culanth.org/curated_collections/11-infrastructure.

Loveband, Anne. 2004. 'Positioning the Product: Indonesian Migrant Women Workers in Taiwan'. *Journal of Contemporary Asia* 34 (3): 336–48. https://doi.org/10.1080/00472330480000141.

Lutz, Helma. 2005. 'Der Privathaushalt Als Weltmarkt Für Weibliche Arbeitskräfte'. *Peripherie* 25 (97/98): 65–87. http://141.2.38.226/www.gesellschaftswissenschaften.uni-frankfurt.de/uploads/7820/2300/Lutz_Weltmarkt-Privathaushalt.pdf.

———. 2008. *Migration and Domestic Work. A European Perspective on a Global Theme*. Aldershot: Ashgate.

———. 2011. *The New Maids: Transnational Women and the Care Economy*. London: Zed Books.

————. 2012. 'At Your Service Madam! The Globalization of Domestic Service'. *Feminist Review* 70 (1) (October): 89–104.

Lutz, Helma, and Ewa Palenga-Möllenbeck. 2010. 'Care Work Migration in Germany: Semi-Compliance and Complicity'. *Social Policy and Society* 9 (3): 419–30. https://doi.org/10.1017/S1474746410000138.

Madden, Janice Fanning. 1981. 'Why Women Work Closer to Home'. *Urban Studies* 18 (2): 181–94. https://doi.org/10.1080/00420988120080341.

Madden, Raymond. 2010. *Being Ethnographic: A Guide to the Theory and Practice of Ethnography*. Thousand Oaks, CA: Sage.

Madörin, Mascha. 2007. 'Neoliberalismus und die Reorganisation der Care-Ökonomie. Eine Forschungsskizze'. *Denknetz Jahrbuch*, 141–62.

————. 2010. 'Care Ökonomie – eine Herausforderung für die Wirtschaftswissenschaften'. In *Gender and Economics. Feministische Kritik der politischen Ökonomie*, edited by Christine Bauhardt and Gülay Çağlar, 81–104. Wiesbaden: VS.

Madörin, Mascha, and Tove Soiland. 2013. 'Care-Ökonomie? Offene Fragen und politische Implikationen'. *Denknetz Jahrbuch*, 84–98.

Maher, Kristen Hill. 2004. 'Good Women "Ready to Go": Labor Brokers and the Transnational Maid Trade'. *Labor* 1 (1): 55–76. https://doi.org/10.1215/15476715-1-1-55.

Mahler, Sarah J., and Patricia R. Pessar. 2001. 'Gendered Geographies of Power: Analyzing Gender Across Transnational Spaces'. *Identities* 7 (4): 441–59. https://doi.org/10.1080/1070289X.2001.9962675.

Marchetti, Sabrina. 2013. 'Dreaming Circularity? Eastern European Women and Job Sharing in Paid Home Care'. *Journal of Immigrant & Refugee Studies* 11 (4): 347–63. https://doi.org/10.1080/15562948.2013.827770.

Massey, Doreen. 1993. 'Power-Geometry and a Progressive Sense of Place'. In *Mapping the Futures: Local Cultures, Global Change*, edited by Jon Bird. Futures, New Perspectives for Cultural Analysis. London; New York, NY: Routledge.

Massey, Douglas S., Joaquin Arango, Graeme Hugo, Ali Kouaouci, Adela Pellegrino, and J. Edward Taylor. 1994. 'An Evaluation of International Migration Theory: The North American Case'. *Population and Development Review* 20 (4): 699–751. www.jstor.org/stable/2137660.

McDowell, Linda. 2009. *Working Bodies: Interactive Service Employment and Workplace Identities*. Studies in Urban and Social Change. Chichester, UK; Malden, MA: Wiley-Blackwell.

McDowell, Linda, Adina Batnitzky, and Sarah Dyer. 2008. 'Internationalization and the Spaces of Temporary Labour: The Global Assembly of a Local Workforce'. *British Journal of Industrial Relations* 46 (4): 750–70. https://doi.org/10.1111/j.1467-8543.2008.00686.x.

McDowell, Linda, Kathryn Ray, Diane Perrons, Colette Fagan, and Kevin Ward. 2011. 'Women's Paid Work and Moral Economies of Care'. *Social and Cultural Geography* 6 (2) (October): 219–35.

Medici, Gabriela. 2012. 'Hauswirtschaft und Betreuung im Privathaushalt. Rechtliche Rahmenbedingungen'. Juristisches Dossier im Auftrag der Fachstelle für Gleichstellung der Stadt Zürich, der Gewerkschaft VPOD und der Gewerkschaft Unia.

————. 2015. *Migrantinnen als Pflegehilfen in Schweizer Privathaushalten: Menschenrechtliche Vorgaben und staatliche Handlungspflichten*. Neue Ausg. Zürcher Studien zum öffentlichen Recht 231. Zürich: Schulthess Juristische Medien.

Medici, Gabriela, and Sarah Schilliger. 2012. 'Arbeitsmarkt Privathaushalt – Pendelmigrantinnen in der Betreuung von alten Menschen'. *Soziale Sicherheit CHSS* 1/2012 (October): 17–20.

Melegh, Attila. 2006. *On the East-West Slope: Globalization, Nationalism, Racism and Discourses on Eastern Europe.* New York, NY: Central European University Press.

Metz-Göckel, Sigrid, Senganata Münst, and Dobrochna Kałwa, eds. 2010. *Migration als Ressource: zur Pendelmigration polnischer Frauen in Privathaushalte der Bundesrepublik.* Opladen: Budrich.

Milligan, Christine, and Janine Wiles. 2010. 'Landscapes of Care'. *Progress in Human Geography* 34 (6): 736–54. https://doi.org/10.1177/0309132510364556.

Misra, Joya, Jonathan Woodring, and Sabine N. Merz. 2006. 'The Globalization of Care Work: Neoliberal Economic Restructuring and Migration Policy'. *Globalizations* 3 (3): 317–32. https://doi.org/10.1080/14747730600870035.

Momsen, Janet Henshall, ed. 1999. *Gender, Migration, and Domestic Service.* London; New York, NY: Routledge.

———. 2012. *Gender, Migration, and Domestic Service.* London: Routledge.

Morokvasic, Mirjana. 1994. 'Pendeln statt Auswandern das Beispiel Polen'. In *Wanderungsraum Europa. Menschen und Grenzen in Bewegung,* 166–87. Berlin: Sigma. www.gesis.org/sowiport/search/id/iz-solis-90194529/.

NAV Hauswirtschaft. 2010a. *NAV Hauswirtschaft: Verordnung über den Normalarbeitsvertrag für Arbeitnehmerinnen und Arbeitsnehmer in der Hauswirtschaft. Stand am 1. Januar 2011.* Vol. 221.215.329.4. www.admin.ch/opc/de/classified-compilation/20102376/index.html.

———. 2010b. *NAV Hauswirtschaft: Verordnung über den Normalarbeitsvertrag für Arbeitnehmerinnen und Arbeitsnehmer in der Hauswirtschaft. Stand am 1. Januar 2014.* Vol. 221.215.329.4. www.admin.ch/opc/de/classified-compilation/20102376/index.html.

NCCR on the move. 2017. 'Projects | NCCR – on the Move'. http://nccr-onthemove.ch/research/projects/.

Novoa, Andre. 2015. 'Mobile Ethnography: Emergence, Techniques and Its Importance to Geography'. *Human Geographies* 9 (1): 97. http://search.proquest.com/openview/4db9b66b11945fefc88da523a06b374a/1?pq-origsite=gscholar&cbl=396324.

OECD. 2011. *Help Wanted? Providing and Paying for Long-Term Care.* Paris: OECD. www.oecd.org/els/health-systems/help-wanted-9789264097759-en.htm.

Österle, August, Andrea Hasl, and Gudrun Bauer. 2013. 'Vermittlungsagenturen in der 24-h-Betreuung'. *WISO. Institut für Sozial- und Wirtschaftswissenschaften* 36 (1): 159–72. www.isw-linz.at/themen/dbdocs/LF_Oesterle_Hasl_Bauer_01_13.pdf.

Pacolet, Jozef, and Frederic De Wispelaere. 2016. *Posting of Workers-Report on A1 Portable Documents Issued in 2015.* Brussels: European Commission. https://lirias.kuleuven.be/bitstream/123456789/573194/1/Posting+of+workers+-+Report+on+A1+portable+documents+issued+in+2015.pdf.

Panayiotopoulos, Prodromos. 2005. 'The Globalisation of Care: Filipina Domestic Workers and Care for the Elderly in Cyprus'. *Capital & Class* 29 (2): 99–134. https://doi.org/10.1177/030981680508600106.

Papadopoulos, Dimitris, and Vassilis Tsianos. 2013. 'After Citizenship: Autonomy of Migration, Organisational Ontology and Mobile Commons'. *Citizenship Studies* 17 (2): 178–96. https://doi.org/10.1080/13621025.2013.780736.

Parrenas, Rhacel Salazar. 2000. 'Migrant Filipina Domestic Workers and the International Division of Reproductive Labor'. *Gender & Society* 14 (4): 560–81. https://doi.org/10.1177/089124300014004005.

———. 2001. *Servants of Globalization: Women, Migration and Domestic Work.* Stanford, CA: Stanford University Press.

Parrenas, Rhacel Salazar, and Rachel Silvey. 2016. 'Domestic Workers Refusing Neo-Slavery in the UAE'. *Contexts* 15 (3): 36–41. https://doi.org/10.1177/1536504216662235.

Paul, Anju Mary. 2013. 'Good Help Is Hard to Find: The Differentiated Mobilisation of Migrant Social Capital Among Filipino Domestic Workers'. *Journal of Ethnic and Migration Studies* 39 (5): 719–39. https://doi.org/10.1080/1369183X.2013.756660.

———. 2017. *Multinational Maids: Stepwise Migration in a Global Labor Market*. Cambridge University Press. Cambridge.

Peck, Jamie, and Nikolas Theodore. 1998. 'The Business of Contingent Work: Growth and Restructuring in Chicago's Temporary Employment Industry'. *Work, Employment and Society* 12 (4): 655–74. https://doi.org/10.1177/0950017098124004.

———. 2001. 'Contingent Chicago: Restructuring the Spaces of Temporary Labor'. *International Journal of Urban and Regional Research* 25 (3): 471–96. https://doi.org/10.1111/1468-2427.00325.

———. 2002. 'Temped Out? Industry Rhetoric, Labor Regulation and Economic Restructuring in the Temporary Staffing Business'. *Economic and Industrial Democracy* 23 (2): 143–75. https://doi.org/10.1177/0143831X02232002.

———. 2006. 'Flexible Recession: The Temporary Staffing Industry and Mediated Work in the United States'. *Cambridge Journal of Economics* 31 (2): 171–92. https://doi.org/10.1093/cje/bel011.

Pelzelmayer, Katharina. 2016. 'Places of Difference: Narratives of Heart-Felt Warmth, Ethnicisation, and Female Care-Migrants in Swiss Live-in Care'. *Gender, Place & Culture* 23 (12): 1701–12. https://doi.org/10.1080/0966369X.2016.1249351.

———. 2017. 'Care, Pay, Love: Commodification and the Spaces of Live-in Care'. *Social & Cultural Geography* 0 (0): 1–20. https://doi.org/10.1080/14649365.2017.1315446.

Perrons, Diane. 2005. 'Gender Mainstreaming and Gender Equality in the New (Market) Economy: An Analysis of Contradictions'. *Social Politics: International Studies in Gender, State & Society* 12 (3): 389–411. https://doi.org/10.1093/sp/jxi021.

Piguet, Etienne. 2004. *Einwanderungsland Schweiz: Fünf Jahrzehnte halb geöffnete Grenzen*. Bern: Haupt.

———. 2006. 'Economy versus the People? Swiss Immigration Policy Between Economic Demand, Xenophobia and International Constraint'. In *Dialogues on Migration Policy*, edited by Marco Giugni and Florence Passy, 67–89. Oxford: Lexington books.

Pijpers, Roos. 2010. 'International Employment Agencies and Migrant Flexiwork in an Enlarged European Union'. *Journal of Ethnic and Migration Studies* 36 (7): 1079–97. https://doi.org/10.1080/13691830903465182.

Piper, Nicola. 2005. 'Rights of Foreign Domestic Workers – Emergence of Transnational and Transregional Solidarity?' *Asian and Pacific Migration Journal* 14 (1–2): 97–119. https://doi.org/10.1177/011719680501400106.

———. 2007. 'Governance of Migration and Transnationalisation of Migrants' Rights. An Organisational Perspective'. http://pub.uni-bielefeld.de/download/2318388/2319944.

———. 2010. 'Temporary Economic Migration and Rights Activism: An Organizational Perspective'. *Ethnic and Racial Studies* 33 (1): 108–25. https://doi.org/10.1080/01419870903023884.

Pratt, Geraldine. 2005. 'From Migrant to Immigrant: Domestic Workers Settle in Vancouver, Canada'. In *A Companion to Feminist Geography*, edited by Lise Nelson and Joni Seager, 123–37. Blackwell Companions to Geography 6. Malden, MA: Blackwell Pub.

———. 2007. 'From Migrant to Immigrant: Domestic Workers Settle in Vancouver, Canada'. In *A Companion to Feminist Geography*, edited by Lise Nelson and Joni Seager, 123–37. Oxford, UK: Blackwell Publishing Ltd. https://doi.org/10.1002/9780470996898.ch9.

————. 2012. *Families Apart. Migrating Mothers and the Conflicts of Labor and Love*. Minneapolis; London: University of Minnesota Press.

Priminfo. 2017. 'Prämienübersicht 2017'. Schweizerische Eidgenossenschaft. Priminfo. www.priminfo.ch/praemien/uebersicht/index.php?sprache=d.

Raghuram, Parvati. 2012. 'Global Care, Local Configurations – Challenges to Conceptualizations of Care'. *Global Networks* 12 (2): 155–74. https://doi.org/10.1111/j.1471-0374.2012.00345.x.

————. 2016. 'Locating Care Ethics Beyond the Global North'. *ACME: An International Journal for Critical Geographies* 15 (3): 511–33. http://oro.open.ac.uk/id/eprint/48887.

Razavi, Shahra. 2007. *The Political and Social Economy of Care in a Development Context: Conceptual Issues, Research Questions and Policy Options*. Gender and Development Paper 3. Geneva: UNRISD.

Rehklau, Christine. 2011. *Die Hausangestelltenfrage in Südafrika zwischen Selbstorganisation und Intervention*. Münster: Waxmann Verlag.

Respekt. 2015. 'Das VPOD-Netzwerk-Respekt für BetreuerInnen in Privathaushalten'. http://respekt-vpod.ch/.

Rigo, Enrica. 2005. 'Citizenship at Europe's Borders: Some Reflections on the Post-Colonial Condition of Europe in the Context of EU Enlargement'. *Citizenship Studies* 9 (1): 3–22. https://doi.org/10.1080/1362102042000325379.

Rippmann, Dorothee. 2011. 'Gesinde. Dienstmädchen im bürgerlichen Haushalt des 19. und 20. Jahrhunderts'. Historisches Lexikon der Schweiz. www.hls-dhs-dss.ch/textes/d/D16376.php.

Rodriguez, Robyn M. 2010. *Migrants for Export. How the Philippine State Brokers Labor to the World*. Minneapolis; London: University of Minnesota Press.

Rodriguez, Robyn M., and Helen Schwenken. 2013. 'Becoming a Migrant at Home: Subjectivation Processes in Migrant-Sending Countries Prior to Departure: Becoming Migrant at Home: Subjectivation Prior to Departure'. *Population, Space and Place* 19 (4): 375–88.

Roseman, Sharon R., Pauline Gardine Barber, and Barbara Neis. 2015. 'Towards A Feminist Political Economy Framework for Analyzing Employment-Related Geographical Mobility'. *Studies in Political Economy* 95 (1): 175–203. https://doi.org/10.1080/1918 7033.2015.11674951.

Rossow, Verena, and Simone Leiber. 2017. 'Zwischen Vermarktlichung und Europäisierung: Die wachsende Bedeutung transnational agierender Vermittlungsagenturen in der häuslichen Pflege in Deutschland'. *Sozialer Fortschritt* 66 (3–4): 285–302. https://doi.org/10.3790/sfo.66.3-4.285.

Rother, Stefan. 2009. 'Transnational Political Spaces: Political Activism of Philippine Labor Migrants in Hong Kong'. In *State, Politics and Nationalism Beyond Borders: Changing Dynamics of Filipino Overseas Migration*, edited by Jorge Tigno, 109–40. Quezon City: Philippine Migration Research Network and the Philippine Social Science Council.

————. 2017. 'Indonesian Migrant Domestic Workers in Transnational Political Spaces: Agency, Gender Roles and Social Class Formation'. *Journal of Ethnic and Migration Studies* (January): 1–18. https://doi.org/10.1080/1369183X.2016.1274567.

Rudnyckyi, Daromir. 2004. 'Technologies of Servitude: Governmentality and Indonesian Transnational Labor Migration'. *Anthropological Quarterly* 77 (3): 407–34. https://doi.org/10.1353/anq.2004.0045.

Ruhs, Martin, and Bridget Anderson. 2006. 'Semi-Compliance in the Migrant Labour Market'. Working Paper No. 30. Centre on Migration, Policy and Society. University of Oxford. http://papers.ssrn.com/sol3/papers.cfm?abstract_id=911280.

Rundschau, S. R. F. 2011. 'Billige Polinnen'. *Beitrag vom 29.6.2011*. www.srf.ch/player/video?id=2c86ba50-7ebf-49d7-8124-09d15b88b46e.

Schälin, Stefanie. 2008. 'Gleichstellungspolitik in der Schweiz'. Seite. Gender Kompetenz Zentrum. www.genderkompetenz.info/genderkompetenz-2003-2010/gendermainstreaming/Strategie/Gleichstellungspolitik/laenderstudien/Schweiz.html.

Schilliger, Sarah. 2009. 'Hauspflege: aktuelle Tendenzen in der Entstehung eines globalisierten, deregulierten Arbeitsmarktes'. *Olympe. Feministische Arbeitshefte zur Politik* 30 (March): 121–26.

———. 2013. 'Transnationale Care-Arbeit: Osteuropäische Pendelmigrantinnen in Privathaushalten von Pflegebedürftigen'. In *Who Cares? Pflege und Solidarität in der alternden Gesellschaft*, edited by Schweizerisches Rotes Kreuz, 142–61. Zürich: Seismo Verlag.

———. 2014. 'Pflegen ohne Grenzen? Polnische Pendelmigrantinnen in der 24h-Betreuung. Eine Ethnographie des Privathaushalts als globalisiertem Arbeitsplatz'. Dissertation, Basel: Universität Basel.

———. 2015. '"Wir sind doch keine Sklavinnen". (Selbst-)Organisierung von polnischen Care-Arbeiterinnen in der Schweiz'. In *Zerstörung und Transformation des Gemeinwesens*. Vol. 8, edited by Hans Baumann, Roland Herzog, Beat Ringger, and Holger Schatz, 164–77. Denknetz Jahrbuch. Zürich.

Schmid-Federer, Barbara. 2012. *Postulat 12.3266 – Rechtliche Rahmenbedingungen für Pendelmigration zur Alterspflege*. www.parlament.ch/de/ratsbetrieb/suche-curia-vista/geschaeft?AffairId=20123266.

Schweizerische Bundeskanzlei. 2017. 'Vorlage Nr. 580'. www.admin.ch/ch/d/pore/va/20140209/det580.html.

Schwenken, Helen. 2013. 'Speedy Latin America, Slow Europe? – Regional Implementation Processes of the ILO Convention on Decent Work for Domestic Workers'. Geneva, Switzerland. www.unrisd.org/80256B3C005BCCF9/(httpAuxPages)/97AA08A7519A3BA9C1257D39005B8205/$file/Schwenken_Regional%20Implementation%20Domestic%20Workers%20Convention.pdf.

Schwiter, Karin, Christian Berndt, and Linda Schilling. 2014. 'Ein sorgender Markt. Wie transnationale Vermittlungsagenturen von Betagtenbetreuerinnen (Im)Mobilität, Ethnizität und Geschlecht in Wert setzen'. *Geographischen Zeitschrift* 102 (4): 2012–231.

Schwiter, Karin, Christian Berndt, and Jasmine Truong. 2018. 'Neoliberal Austerity and the Marketisation of Elderly Care'. *Social & Cultural Geography* 19 (3): 379–99. https://doi.org/10.1080/14649365.2015.1059473.

Schwiter, Karin, Katharina Pelzelmayer, and Isabelle Thurnherr. 2018. 'On the Construction of 24 Hours Care for the Elderly in the Swiss Media'. *Swiss Journal of Sociology* 44 (1): 157–81. https://doi.org/10.1515/sjs-2018-0008.

———. Forthcoming. 'Ein boomender Markt? Die Konstruktion der 24h-Stunden-Betreuung für ältere Menschen in den Schweizer Medien'. *Swiss Journal of Sociology*.

SECO, Swiss State Secretariat for Economic Affairs. 2012. Rechtliche Rahmenbedingungen für Pendelmigration zur Alterspflege'. *Bericht des Bundesrates in Erfüllung des Postulats Schmid-Federer 12.3266 vom 16. März 2012.*

———. 2017. 'Entsendung Und Flankierende Massnahmen'. www.seco.admin.ch/seco/de/home/Arbeit/Personenfreizuegigkeit_Arbeitsbeziehungen/freier-personenverkehr-ch-eu-und-flankierende-massnahmen.html.

SFSO, Swiss Federal Statistical Office. 2010. *Szenarien zur Bevölkerungsentwicklung der Schweiz 2010–2016*. Neuchâtel: Swiss Federal Statistical Office. www.bfs.admin.ch/

bfs/de/home/statistiken/arbeit-erwerb/erwerbstaetigkeit-arbeitszeit/erwerbspersonen/
szenarien-erwerbsbevoelkerung.assetdetail.350324.html.

———. 2014a. 'Hauptverantwortung für Hausarbeiten'. Schweizerische Arbeitskräfteerhebung (SAKE): Modul Unbezahlte Arbeit. www.bfs.admin.ch/bfs/de/home/statistiken/
wirtschaftliche-soziale-situation-bevoelkerung/gleichstellung-frau-mann/vereinbarkeit-
beruf-familie/hauptverantwortung-hausarbeiten.html.

———. 2014b. 'Hausarbeit und Familienarbeit'. Schweizerische Arbeitskräfteerhebung
(SAKE): Modul Unbezahlte Arbeit. www.bfs.admin.ch/bfs/de/home/statistiken/arbeit-
erwerb/unbezahlte-arbeit/haus-familienarbeit.html.

———. 2016. *Ein Portrait der Schweiz: Ergebnisse aus den Volkszählungen 2010–2014*.
Neuchâtel: Bundesamt für Statistik BFS.

Sheller, Mimi. 2016. 'Uneven Mobility Futures: A Foucauldian Approach'. *Mobilities* 11
(1): 15–31. https://doi.org/10.1080/17450101.2015.1097038.

Sheller, Mimi, and John Urry. 2006. 'The New Mobilities Paradigm'. *Environment and
Planning A* 38 (2): 207–26. https://doi.org/10.1068/a37268.

Shubin, Sergei, Allan Findlay, and David McCollum. 2014. 'Imaginaries of the Ideal
Migrant Worker: A Lacanian Interpretation'. *Environment and Planning D: Society and
Space* 32 (3): 466–83. www.envplan.com/abstract.cgi?id=d22212.

Shutes, Isabel, and Carlos Chiatti. 2012. 'Migrant Labour and the Marketisation of
Care for Older People: The Employment of Migrant Care Workers by Families and
Service Providers'. *Journal of European Social Policy* 22 (4): 392–405. https://doi.
org/10.1177/0958928712449773.

Silvey, Rachel. 2004. 'Transnational Domestication: State Power and Indonesian Migrant
Women in Saudi Arabia'. *Political Geography* 23 (3): 245–64. https://doi.org/10.1016/j.
polgeo.2003.12.015.

Smith, Michael Peter. 2005. 'Transnational Urbanism Revisited'. *Journal of Ethnic and
Migration Studies* 31 (2): 235–44. https://doi.org/10.1080/1369183042000339909.

SOCZ, Statistical Office of the Canton of Zurich. 2008. *Babyboomer Kommen Ins Rentan-
alter. Der Lebenszyklus Der Geburtenstarken Jahrgänge Im Kanton Zürich 1970–2050*.
Zürich: Statistical Office of the Canton of Zurich. www.statistik.zh.ch/dam/justiz_
innern/statistik/Publikationen/statistik_info/si_2008_06_babayboomer.pdf.spooler.
download.1326986364531.pdf/si_2008_06_babayboomer.pdf.

SRF, Schweizer Radio und Fernsehen. 2013a. 'McCare wegen Billigbetreuung in der Kritik'.
10vor10. Sendung vom 02.05.2013. www.srf.ch/player/tv/10vor10/video/mccare-wegen-
billigbetreuung-in-der-kritik?id=01aa8ced-e4d0-4101-8480-89af9d1bd449.

———. 2013b. 'Sieben Tage Arbeit, fünf Tage Lohn'. *10vor10. Sendung vom 01.05.2013*.
www.srf.ch/player/tv/10vor10/video/sieben-tage-arbeit-fuenf-tage-lohn?id=e3016a42-
7a4a-4bd8-9d08-490eb34b5314.

———. 2013c. 'Gesucht: Pflegerin aus dem Osten'. *Club. Sendung vom 13.08.2013*. www.
srf.ch/player/video?id=55F4F856-4F37-4FE9-85E4-608DDCA8259E&referrer=http%
253A%252F%252Ftvprogramm.srf.ch%252F13.08.2013.

Star, Susan Leigh. 1999. 'The Ethnography of Infrastructure'. *American Behavioral Scientist* 43 (3): 377–91. https://doi.org/10.1177/00027649921955326.

Steinberg, Ronnie J., and Deborah M. Figart. 1999. 'Emotional Labor Since: The Managed
Heart'. *The Annals of the American Academy of Political and Social Science* 561 (1):
8–26. https://doi.org/10.1177/000271629956100101.

Stephen, Castles, and Mark J. Miller. 2009. *The Age of Migration*. 4th ed. Basingstoke:
Palgrave Macmillan.

Stiell, Bernadette, and Kim England. 1999. 'Jamaican Domestics, Filipina Housekeepers and English Nannies: Representations of Toronto's Foreign Domestic Workers'. In *Gender, Migration and Domestic Service*, edited by Janet Henshall Momsen, 43–60. London; New York, NY: Routledge.

Strauss, Anselm, and Juliet Corbin. 1990. *Basics of Qualitative Research. Techniques and Procedures for Developing Theory.* 2nd ed. Vol. 15. Newbury Park: Sage.

Strüver, Anke. 2011. 'Zwischen Care Und Career – Haushaltsnahe Dienstleistungen von Transnational Mobilen Migrantinnen Als Strategische Ressourcen'. *Zeitschrift Für Wirtschaftsgeographie* 55 (4) (October): 193–206. https://doi.org/10.1515/zfw.2011.0015.

———. 2013. 'Ich war lange illegal hier, aber jetzt hat mich die Grenze übertreten – Subjektivierungsprozesse transnational mobiler Haushaltshilfen'. *Geographica Helvetica* 68 (3): 191–200. https://doi.org/10.5194/gh-68-191-2013.

Transparency Market Research. 2016. 'Aged Care Market – Industry Analysis, Share, Forecast 2023'. www.transparencymarketresearch.com/aged-care-market.html.

Triandafyllidou, Anna. 2010. *METOIKOS Project. Towards a Better Understanding of Circular Migration.* Florence: European University Institute, Robert Schuman Centre for Advanced Studies. www.eui.eu/Projects/METOIKOS/Documents/ConceptPaper/METOIKOSConceptPaper1July2010.pdf.

Triandafyllidou, Anna, and Sabrina Marchetti. 2013. 'Migrant Domestic and Care Workers in Europe: New Patterns of Circulation?' *Journal of Immigrant & Refugee Studies* 11 (4): 339–46. https://doi.org/10.1080/15562948.2013.822750.

Truong, Jasmine. 2011. 'Arbeit, Arbeitsidentität, Arbeitsplatz. Die neuen Wanderarbeiterinnen in der Sorgewirtschaft'. Masterarbeit, Universität Zürich.

Truong, Jasmine, Karin Schwiter, and Christian Berndt. 2012. *Arbeitsmarkt Privathaushalt. Charakteristika der Unternehmen, deren Beschäftigungsstruktur und Arbeitsbedingungen. Eine Studie im Auftrag der Fachstelle für Gleichstellung der Stadt Zürich.* Zürich: Geographisches Institut der Universität Zürich.

UNIA. 2014. 'Gute Arbeitsbedingungen in Der Privaten Altersbetreuung'. www.unia.ch/de/medien/medienmitteilungen/mitteilung/a/9847/.

UNIA & Zu Hause Leben. 2014. *Gesamtarbeitsvertrag für den Bereich private nicht medizinische Betreuung in der Deutschschweiz.* www.unia.ch/de/medien/medienmitteilungen/mitteilung/a/9847/.

United Nations, Department of Economic and Social Affairs, Population Division. 2015. *World Population Ageing 2015.* ST/ESA/SER.A/390. New York. www.un.org/en/development/desa/population/publications/pdf/ageing/WPA2015_Report.pdf.

Urry, John. 2003. *Global Complexity.* Cambridge: Polity Press.

———. 2007. *Mobilities.* Cambridge: Polity Press.

Van Holten, Karin, Anke Jähnke, and Iren Bischofberger. 2013. *Care-Migration – Transnationale Sorgearrangements im Privathaushalt.* 57. Obsan Bericht. Neuchâtel: Schweizerisches Gesundheitsobservatorium. www.obsan.admin.ch/sites/default/files/publications/2015/obsan_57_bericht.pdf.

Van Holten, Karin, Anke Jähnke, Iren Bischofberger, and Schweizerisches Gesundheitsobservatorium. 2013. *Care-Migration-Transnationale Sorgearrangements Im Privathaushalt.* Schweizerisches Gesundheitsobservatorium. www.bfs.admin.ch/bfs/portal/fr/index/themen/14/22/publ.Document.170657.pdf.

Vertovec, Steven. 2007. 'Circular Migration: The Way Forward in Global Policy'. International Migration Institute (IMI) Working Papers, IMI, University of Oxford. www.imi.ox.ac.uk/pdfs/imi-working-papers/wp4-circular-migration-policy.pdf.

Ward, Kevin. 2004. 'Going Global? Internationalization and Diversification in the Temporary Staffing Industry'. *Journal of Economic Geography* 4 (3): 251–73. https://doi.org/10.1093/jnlecg/lbh019.

Waring, Marilyn, and Gloria Steinem. 1988. *If Women Counted: A New Feminist Economics*. London: Macmillan.

Wee. 2019. 'Brokering Migrant Domestic Workers in Singapore. Maid Agents and the Puzzle of Moral Credibility'. Master thesis, Singapore: National University of Singapore.

Wehrli, Markus. 2013. 'Billigpflege kann teuer werden'. *St. Galler Tagblatt*, 5 March 2013.

Wicker, Hans-Rudolf. 2010. 'Deportation at the Limits of "Tolerance". The Juridical, Institutional, and Social Construction of "Illegality" in Switzerland'. In *The Deportation Regime. Sovereignty, Space, and the Freedom of Movement*, edited by Nicholas De Genova and Nathalie Peutz, 224–44. Durham; London: Duke University Press.

Wickramasekara, Piyasiri. 2011. 'Circular Migration: A Triple Win or a Dead End?' *International Labour Office. Bureau for Workers' Activities* 15: 114. https://doi.org/10.2139/ssrn.1834762.

Wigger, Annegret, Nadia Baghdadi, and Bettina Brüschweiler. 2013. '"Care"-Trends in Privathaushalten: Umverteilen oder auslagern?' In *Who Cares? Pflege und Solidarität in der alternden Gesellschaft*, edited by Schweizerisches Rotes Kreuz, 82–103. Zürich: Seismo. www.redcross.ch/data/info/pubs/pdf/redcross_584_de.pdf.

Williams, Fiona. 2011. 'Markets and Migrants in the Care Economy'. *Soundings* 47: 22–33. https://doi.org/10.3898/136266211795427576.

Wimmer, Andreas, and Nina Glick Schiller. 2002. 'Methodological Nationalism and Beyond: Nation – State Building, Migration and the Social Sciences'. *Global Networks* 2 (4): 301–34. https://doi.org/10.1111/1471-0374.00043.

Winker, Gabriele. 2009. 'Care Revolution – ein Weg aus der Reproduktionskrise'. Feministisches Institut Hamburg. www.feministisches-institut.de/wp-content/uploads/2009/12/CareRevolution.pdf.

———. 2015. *Care Revolution: Schritte in eine solidarische Gesellschaft*. X-Texte. Bielefeld: Trascript.

'XE Currency Converter'. 2017. www.xe.com/de/currencytables/?from=CHF&date=2014-05-15.

Xiang, Biao, and Johan Lindquist. 2014. 'Migration Infrastructure'. *International Migration Review* 48: 122–48. https://doi.org/10.1111/imre.12141.

Yeates, Nicola. 2004. 'Global Care Chains'. *International Feminist Journal of Politics* 6 (3): 369–91. https://doi.org/10.1080/1461674042000235573.

———. 2012. 'Global Care Chains: A State-of-the-Art Review and Future Directions in Care Transnationalization Research'. *Global Networks* 12 (2): 135–54. https://doi.org/10.1111/j.1471-0374.2012.00344.x.

Yeoh, Brenda S. A., and Shirlena Huang. 2010. 'Transnational Domestic Workers and the Negotiation of Mobility and Work Practices in Singapore's Home-Spaces'. *Mobilities* 5 (2): 219–36.

Yeoh, Brenda S. A., Shirlena Huang, and Theresa W. Devasahayam. 2004. 'Diasporic Subjects in the Nation: Foreign Domestic Workers, the Reach of Law and Civil Society in Singapore'. *Asian Studies Review* 28 (1): 7–23. https://doi.org/10.1080/1035782042000194491.

Index

Printed in the United States
by Baker & Taylor Publisher Services